T0220312

Cambridge Elements ≡

Elements of Aerospace Engineering
edited by
Vigor Yang
Georgia Institute of Technology
Wei Shyy
Hong Kong University of Science and Technology

MACH WAVE AND ACOUSTICAL WAVE STRUCTURE IN NONEQUILIBRIUM GAS-PARTICLE FLOWS

Joseph T. C. Liu

Brown University

CAMBRIDGE
UNIVERSITY PRESS

CAMBRIDGE
UNIVERSITY PRESS

University Printing House, Cambridge CB2 8BS, United Kingdom

One Liberty Plaza, 20th Floor, New York, NY 10006, USA

477 Williamstown Road, Port Melbourne, VIC 3207, Australia

314–321, 3rd Floor, Plot 3, Splendor Forum, Jasola District Centre,
New Delhi – 110025, India

103 Penang Road, #05–06/07, Visioncrest Commercial, Singapore 238467

Cambridge University Press is part of the University of Cambridge.

It furthers the University's mission by disseminating knowledge in the pursuit of
education, learning, and research at the highest international levels of excellence.

www.cambridge.org
Information on this title: www.cambridge.org/9781108964883
DOI: 10.1017/9781108990585

© Joseph T. C. Liu 2021

First published 2021

A catalogue record for this publication is available from the British Library.

ISBN 978-1-108-96488-3 Paperback
ISSN 2631-7850 (online)
ISSN 2631-7842 (print)

Mach Wave and Acoustical Wave Structure in Nonequilibrium Gas-Particle Flows

Elements of Aerospace Engineering

DOI: 10.1017/9781108990585
First published online: September 2021

Joseph T. C. Liu
Brown University

Author for correspondence: Joseph T. C. Liu, joseph_liu@brown.edu

Abstract: In this Element, the gas-particle flow problem is formulated with momentum and thermal slip that introduces two relaxation times. Starting from acoustical propagation in a medium in equilibrium, the relaxation-wave equation in airfoil coordinates is derived though a Galilean transformation for uniform flow. Steady planar small perturbation supersonic flow is studied in detail according to Whitham's higher-order waves. The signals owing to wall boundary conditions are damped along the frozen-Mach wave, and are both damped and diffusive along an effective-intermediate Mach wave and diffusive along the equilibrium Mach wave where the bulk of the disturbance propagates. The surface pressure coefficient is obtained exactly for small-disturbance theory, but it is considerably simplified for the small particle-to-gas mass loading approximation, equivalent to a simple-wave approximation. Other relaxation-wave problems are discussed. Martian dust-storm properties in terms of gas-particle flow parameters are estimated.

Keywords: Mach waves, gas-particle flow, wall pressure coefficient, analytical methods, acoustic waves

ISBNs: 9781108964883 (PB), 9781108990585 (OC)
ISSNs: 2631-7850 (online), 2631-7842 (print)

To
Shôn E. Ffowcs Williams
Sau-Hai (Harvey) Lam
Lester Lees
M. James Lighthill
Frank E. Marble
Ronald F. Probstein
J. Trevor Stuart
with respect and gratitude

Contents

Contents

1 Introduction

The motion of a fluid containing small solid particles has been of interest in many scientific and engineering disciplines since the turn of the previous century. In a paper on sand ripples in the desert, Theodore von Kármán (1947a) gave an invitation and pointed out the challenges to workers in fluid mechanics in elucidating problems in this field. Recent issues of aerospace engineering interest concern the aerodynamics of dusty planetary atmospheres. Early work recognized the presence of micron-sized dust particles in the atmospheres of Mars and Venus (Öpik 1962). More recent work on Martian atmospheric dust properties is presented in Appendix B; like the present continuum considerations of gas-particle flow, these are of current importance in planning human exploration of Mars (Levine and Winterhalter 2017).

Reviews of earlier, largely empirical works (Dalla Valle 1948; Hermans 1953; Othmer 1956; Torobin and Gauvin 1959) on such problems as atomization of liquids, fluidization, powder beds, and smoke and raindrop impingement and icing on airfoils (Serafini 1954; Gelder et al. 1956; Lewis and Brun 1956) are mainly concerned with trajectories of single droplets in an undisturbed aerodynamic flow field. The subjects subsumed by "heterogeneous flows" are sufficiently numerous that many books on the subjects have emerged – for example, Soo (1967), Rudinger (1980), Fan and Zhu (1998), Jackson (2000), Brenan (2005), Guazzelli and Morris (2012), and Michaelides (2014) to mention just a few.

The aeronautical and aerospace engineering interests in gas-particle flows stem from combustion instabilities in solid propellant rocket motors (Culick and Yang 1992) and their alleviation via embedded micron-sized aluminum oxide particles that damp out acoustical instabilities. Here, a host of gas-particle flow problems of aeronautical interest arise. Particle trajectories in a cascade flow were studied by Tabakoff and Hussein (1971) in a wind tunnel. The review by Hoglund (1962) of gas-particle flow in converging-diverging nozzles was largely taken up with tedious numerical procedures. Analytical descriptions of the nozzle problem were given by Rannie (1962) and Marble (1963). The normal shock wave in a gas containing small solid particles was first studied by Carrier (1958) and by Marble (1962), and it was also described in Rudinger (1964, 1980), Marble (1970), and Brenan (2005) and in reviews (e.g., Ingra and Ben-Dor 1988). The oblique shock wave on a wedge was studied by Miura and Glass (1986). Particle trajectories in a Prandtl–Meyer expansion gas flow were studied by Marble (1962) and in a gas-particle flow by Miura and Glass (1988). Work on weak waves over slender wedges was studied by Liu (1964) from a general small-disturbance theory point of view, and by Miura (1974). It seems that the relaxation wave equation was overlooked in Miura (1974).

The hydrodynamic stability of parallel laminar gas-particle flow was studied by Saffman (1962) and Michael (1964, 1965). Liu (1965) studied how the Stokes layer near the wall, which contributes to the Reynolds stress, is subject to frequency-dependent modification in gas-particle flow.

The discussion in the present work is necessarily modest and covers only a small segment of the vast subject of gas-particle flows. We deal with small solid particles in a gaseous medium, a "dusty gas," in which the small solid particles occupy negligible volume in a unit volume of the mixture. The solid particulate material density is ρ_S and the particle density per unit volume of the mixture is $\rho_P = \rho_S n_P$. Our range of interest is $\kappa = \rho_P/\rho = \vartheta(1)$, where the particulate phase momentum and thermal lag have an impact on the gas flow. When $\kappa \to 0$, the gas flow is not affected and the central interest is the study of particle trajectories in a known flow field (Glauert 1940; Probstein and Fassio 1970).

General wave problems in fluids are presented in Whitham (1974) and in Lighthill (1978).

2 Conservation Equations of Gas-Particle Flows

The motion of a gas containing small solid particles (a dusty gas) was discussed in rather general terms by Kiely (1959); he stated the equations of motion and applied the techniques of irreversible processes for small departures from thermodynamic equilibrium to deduce the forms of the particle-fluid interaction force. However, he did not point out the dissipative mechanism resulting from the work done owing to particle-fluid momentum interaction, and this omission is perpetuated in the subsequent calculation of entropy sources. This was remedied by Chu and Parlange (1962), also from an irreversible thermodynamic point of view (Prigogine 1961), who obtained laws for momentum and thermal interaction between the two phases. As expected, these laws are linear in the velocity and temperature differences, which are the forms that Stokes's law takes. In this context, the linear interaction laws are on the same footing as the Newtonian and Fourier linear relations between fluxes of momentum and of heat and the gradients of velocity and temperature, respectively, for small departures from thermo-dynamic equilibrium.

The conservation equations were also discussed by Van Deemter and Van der Laan (1961) and Hinze (1962), but only for mass, momentum, and kinetic energy. The omission of thermodynamic considerations of thermal energy renders them incomplete. Then Marble discussed the general continuum conservation equations on the basis of particle distribution function (Marble 1962), and the effect of phase change (Marble 1969). In the present work, we discuss in detail inert particles with only momentum and thermal interactions between the two phases. Consider metallic solid particles for which the ratio of mass density of the solid material to that of a gas at standard conditions is estimated to be $\rho_S/\rho = \vartheta(10^3)$. Our interest is in the range in which the mass density of the solid phase $\kappa = \rho_P/\rho = n_P \rho_S/\rho = \vartheta(1)$, for which there is mutual influence between the two phases. In this case, the total volume occupied by the solid phase in a unit volume of the mixture is $\vartheta(10^{-3})$. Thus the per-unit volume of the mixture is synonymous with the per-unit volume of the gas due to the negligible volume occupied by the solid particles. Furthermore, for $\kappa = \vartheta(1)$, if the average particle radius is $r_P = 1, 10$ *micron*, then $n_P = \vartheta(10^5)$ and the interparticle distance would be about 10^{-2} *mm*. For $r_P = 10$ *micron*, these would be, respectively, $n_P = \vartheta(10^3)$ and 10^{-1} *mm*. Thus one can certainly define a macroscopic "point" the size of a fraction of a millimeter over which an average quantity of particle cloud can be suitably defined. Concurrently, the interparticle distance is sufficiently large compared to the size of the particles so as to render particle-particle interactions secondary compared to particle-fluid interactions. For particles of disparate size, "collisions" become unavoidable so that this situation is a separate consideration (Marble 1970). Here, we consider only particles of a single averaged size. The particle-fluid interactions are continuous since the mean free path of the gaseous medium is of the order of 5×10^{-5} *mm* and the ratio of the mass of a single particle to that of a gas molecule is of the order of 10^{11} and 10^{14} for $r_P = 1$ and 10 micron, respectively.

We thus consider a perfect gas coexisting with small solid particles that we assume to be spherical and sufficiently rare and non-mutually interacting. Within a macroscopic point, in general, the individual particles may have different velocities, directions, and temperatures. Owing to the large number of particles within the macroscopic point, averaged local properties such as density, velocity, and temperature are defined, and these are point functions as in ordinary gas dynamics. The conservation equations are:

Conservation of mass – In the absence of mass exchange between the two phases, the continuity equation for the gas phase is:

$$\frac{\partial \rho}{\partial t} + \frac{\partial (\rho u_j)}{\partial x_j} = 0 \tag{2.1}$$

and for the particle phase, it is:

$$\frac{\partial \rho_P}{\partial t} + \frac{\partial(\rho_P u_{Pj})}{\partial x_j} = 0 \tag{2.2}.$$

Conservation of momentum – The momentum equation for the gas phase takes the usual Navier–Stokes form, but is augmented by the force per unit volume exerted on the gas by the particle phase F_{Pi}:

$$\frac{\partial(\rho u_i)}{\partial t} + \frac{\partial(\rho u_j u_i)}{\partial x_j} = -\frac{\partial p}{\partial x_i} + \frac{\partial \tau_{ji}}{\partial x_j} + F_{Pi} \tag{2.3}.$$

The usual viscous stress tensor τ_{ji}, which is linearly related to the gas phase rate of strain tensor, refers to smooth derivatives of the gas velocities. In this case, the detailed disturbance on the gas velocity due to passage of the particles is neglected.

The particle phase momentum equation is:

$$\frac{\partial(\rho_P u_{Pi})}{\partial t} + \frac{\partial(\rho_P u_{Pj} u_{Pi})}{\partial x_j} = -F_{Pi} \tag{2.4},$$

which makes no contribution to its own pressure and viscous stresses according to our discussions (Marble 1970). The force per unit volume exerted on the particle phase by the gas is $-F_{Pi}$.

Conservation of energy – The energy equation for the gas phase in the form of the first law of thermodynamics for the internal energy per unit mass is:

$$\frac{\partial(\rho e)}{\partial t} + \frac{\partial(\rho u_j e)}{\partial x_j} + p\frac{\partial u_j}{\partial x_j} = -\frac{\partial q_j}{\partial x_j} + \Phi + Q_P + \Phi_P \tag{2.5},$$

where Q_P is the rate of heat transferred per unit mass from the particle phase to the gas phase, Φ_P is the dissipative mechanism owing to the work done on the gas by the particle phase and is $(u_{Pi} - u_i)F_{Pi}$, q_i is the Fourier heat flux vector (again, it refers to smooth derivatives of the gas temperature in the Fourier assumption), and Φ is the rate of viscous dissipation. The particle phase energy equation, in terms of the particle phase internal energy per unit mass, $e_p = c_s T_P$, is:

$$\frac{\partial(\rho_p e_P)}{\partial t} + \frac{\partial(\rho_P u_{Pj} e_P)}{\partial x_j} = -Q_P \tag{2.6}.$$

The rate at which heat is transferred per unit volume from the gas to the particle phase is $(-Q_P)$.

Equation of state – The gas phase for moderate temperatures and pressures is assumed to be a perfect gas, satisfying

$$p = \rho RT \tag{2.7}.$$

The particle phase is not constrained by an equation of state owing to its lack of volume and randomizing effects.

Particle phase-gas phase interaction force and heat transfer – Irreversible thermodynamic considerations (Prigogine 1961; Chu and Parlange 1962) provide the interactions in the form of linear differences between the velocities and temperatures for momentum and heat transfer, respectively, between the phases. The "coefficients" of such differences are a separate matter, much like the viscosity and thermal conductivity in the linear relation between stress and rates of strain and between the heat flux and the temperature gradient. A way of evaluating the "coefficients" is Stokes's law, which has already yielded the linear differences.

Since the particle phase occupies negligible volume, the force exerted on the particle phase due to the pressure gradient in the gas, as well as that due to virtual mass, is neglected. The interaction force per unit volume for the particle phase-gas phase interaction for a linear relation in the relative velocity is:

$$F_{Pi} = \frac{\rho_P}{\tau_V}(u_{Pi} - u_i) \tag{2.8}.$$

If we assume that Stokes's law holds, then the relation gives the velocity relaxation time as

$$\tau_V = m_P/(6\pi\mu r_P) \tag{2.9}.$$

The particles are assumed noninteracting in the unit volume of the mixture, as discussed earlier in this Element. Similarly, the heat transferred to the particle phase is represented as a linear dependence on the relative temperature difference between the two phases:

$$Q_P = \frac{\rho_P}{\tau_T}c_S(T_P - T) \tag{2.10},$$

where the thermal relaxation time for particles obeying Stokes's law is:

$$\tau_T = \frac{m_P c_S}{4\pi r_P k} \tag{2.11}.$$

For metallic particles, the two relaxation times are of the same order.

It has been tacitly assumed, since the interparticle distance in our range of interest is much larger than the particle size, that the interaction between the particle phase in a unit volume of mixture is equal to the number of particles in that volume times the corresponding effect of a single particle.

Rubinow and Keller (1961) showed that the transverse force on a sphere in shear flow is solely due to its spin. However, when a sphere originally lacks spin, it continues to lack it thereafter. This is assumed to be the case, and the transverse force is not included in our considerations.

2.1 Incompressible Gas Flow

The reduction of the gas-phase conservation equations to an incompressible form generally follows from Lagerstrom (1996), where flow quantities undergo a double expansion for small-temperature loading and small Mach numbers. The former expansion renders negligible the variation in gas and transport properties; the latter expansion makes the rate of viscous dissipation and the work done by the pressure gradients unimportant.

For gas-particle flows, additionally the relative temperature differences between the phases and the Mach number based on the relative velocities are small. The first restriction is to keep the gas density and transport properties relatively constant; the second is to render negligible the heat source in the gas due to work done on the gas by the force exerted by the particle phase. Since the particle phase is not constrained by an equation of state, such as those for an ideal gas for the gas phase, the particle density per unit volume of the mixture, ρ_P, is governed solely by mass conservation considerations and the particle-phase continuity equation does not have an incompressible form.

For incompressible gas-particle flows, then,

$$\frac{\partial u_j}{\partial x_j} = 0$$
$$\frac{\partial u_i}{\partial t} + u_j \frac{\partial u_i}{\partial x_j} = -\frac{1}{\rho_0}\frac{\partial p}{\partial x_i} + v_0 \frac{\partial}{\partial x_j}\left(\frac{\partial u_i}{\partial x_j} + \frac{\partial u_j}{\partial x_i}\right) + \frac{\rho_P}{\rho_0}\frac{1}{\tau_{V0}}(u_{Pi} - u_i)$$

(2.12),

where the Navier–Stokes relation for the viscous stress tensor has been used and the subscript 0 denotes free-stream conditions, $\tau_{V0} = m_P/6\pi\mu_0 r_P$.

The corresponding particle phase continuity and momentum equations become:

$$\frac{\partial \rho_P}{\partial t} + \frac{\partial(\rho_P u_{Pj})}{\partial x_j} = 0$$
$$\frac{\partial(\rho_P u_{Pi})}{\partial t} + \frac{\partial(\rho_P u_{Pj} u_{Pi})}{\partial x_j} = -\frac{\rho_P}{\tau_{V0}}(u_{Pi} - u_i)$$

(2.13).

The "incompressible" form of the gas-particle flow equations forms the basis of boundary-layer studies (Saffman 1962; Marble 1962, 1970; Michael 1964; Liu 1966, 1967). Viscous boundary layer problems are essentially "transverse wave" problems in which the diffusion process relaxes from frozen diffusion via $\nu = \mu/\rho$, to equilibrium diffusion via $\nu_e = \mu/(1+\kappa)\rho$.

The thermal problems are uncoupled from the above, in which the momentum problem is considered solved and used as input in the thermodynamic problem. The energy equations become, for the heat transfer problem,

$$
\frac{\partial T}{\partial t} + u_j \frac{\partial T}{\partial x_j} = K_0 \frac{\partial^2 T}{\partial x_j^2} + \frac{\rho_P c_S}{\rho_0 c_P} \frac{1}{\tau_{T0}}(T_P - T)
$$
$$
\frac{\partial T_P}{\partial t} + u_j \frac{\partial T_P}{\partial x_j} = -\rho_P c_S \frac{1}{\tau_{T0}}(T_P - T)
$$

(2.14),

where $\tau_{T0} = m_P c_S / 4\pi k_0 r_P$ and the thermal diffusivity is $K_0 = k_0/\rho_0 c_P$. For the insulated wall problem or the thermometer problem, dissipative effects would be included.

2.2 Inviscid Compressible Flows

In gas-particle flow fields of aeronautical interest, those resulting from the obstacles in the stream, the classification of certain regions in the flow field is similar to that in gas dynamics for gases with negligible friction, since only transport effects of the gas are considered. In other words, shearing stresses and heat conduction are confined to thin boundary layers adjacent to solid boundaries when the flow Reynolds number, based on some macroscopic characteristic length, is large. Outside such regions, the gas-particle momentum interaction that arises due to gas viscosity is more important compared to the viscous forces acting on a solid boundary, and similarly for other transport effects such as heat conductivity and mass diffusion. Loosely speaking, then, outside certain thin boundary layer regions, we may consider the "inviscid" flow in gas-particle flows.

For the gas phase, which can be compressible, the conservation equations take the form:

$$
\frac{\partial \rho}{\partial t} + \frac{\partial (\rho u_i)}{\partial x_j} = 0
$$
$$
\frac{\partial (\rho u_i)}{\partial t} + \frac{\partial (\rho u_j u_i)}{\partial x_j} = -\frac{\partial p}{\partial x_i} + F_{P_i}
$$
$$
\frac{\partial (\rho e)}{\partial t} + \frac{\partial (\rho u_j e)}{\partial x_j} + p \frac{\partial u_j}{\partial x_j} = Q_P + (u_{Pi} - u_i)F_{Pi}
$$
$$
p = \rho R T
$$

(2.15).

In terms of the gas static enthalpy per unit mass, $h = e + p/\rho$, the more convenient energy equation for flow problems is:

$$\frac{\partial(\rho h)}{\partial t} + \frac{\partial(\rho u_j h)}{\partial x_j} = \frac{\partial p}{\partial t} + u_j \frac{\partial p}{\partial x_j} + Q_P + (u_{Pi} - u_i)F_{Pi} \qquad (2.16).$$

The corresponding conservation equation for the particle phase remains the same as (2.2), (2.4), and (2.6). The enthalpy of the particle phase is the same as its internal energy, $h_P = c_S T_P = e_P$.

The entropy (per unit mass) equations for the two phases are, following standard definitions,

$$\rho \frac{\partial s}{\partial t} + \rho u_j \frac{\partial s}{\partial x_j} = \frac{1}{T}[Q_P + (u_{Pi} - u_i)F_{Pi}]$$
$$\rho_P \frac{\partial s_P}{\partial t} + \rho_P u_{Pj} \frac{\partial s_P}{\partial x_j} = \frac{1}{T_P}[-Q_P] \qquad (2.17).$$

The pioneering work of Ackeret (1925), Glauert (1928), Kármán and Moore (1932), and Kármán (1941) and the lectures of Busemann (1935), Prandtl (1935), and Kármán (1935) at the Fifth Volta Congress for High Speed Aeronautics held in 1935 in Rome essentially established the foundations of small disturbance theory in supersonic flows. One only need to consult the two subsequent general lectures of Kármán, the Tenth Wright Brothers Lectures of 1947 (Kármán 1947b), and the Fifth Guggenheim Memorial Lecture of 1958 (Kármán 1959) to gain perspective of the developments of supersonic flow. Sears (1954), among others, connected acoustics (Rayleigh [1894] 1945) with aerodynamics through the assumption that the thin body emits acoustic waves as it travels, transforming the acoustic wave equation into the aerodynamics of unsteady supersonic flow (Miles 1959). What is more, in the special case of steady flow, the Prandtl–Glauert–Ackeret equation emerges. These ideas are followed here from acoustic wave propagation into steady supersonic flow over thin bodies in gas-particle flow.

The inviscid, compressible conservation equations are given in Section 2. They are the "Euler equations" in gas-particle flows that are to be linearized, leading to the formulation of small-disturbance perturbation theory. In gas dynamics, the development of acoustical theory, in terms of small perturbations about a stationary medium (Rayleigh [1894] 1945), is unified with aerodynamics of slender or thin bodies, in small perturbations from a uniform stream, the Prandtl–Glauert perturbation equation, as reviewed by Sears (1954) concerning the historical evolution of small perturbation theory in gas dynamics. The physical implication of linearized thin airfoil theory in a compressible gas is that the flow is governed by the equations of acoustics in the frame of reference fixed on the airfoil and where the perturbation of the gas velocities is assumed to be small compared to the acoustic propagation speed. This general idea is adopted in what follows, proceeding from acoustic propagation in a gas-particle medium originally at rest and in thermal equilibrium (Chu and Parlange 1962). The Prandtl–Glauert–Ackeret form of the equation in gas-particle flow (obtained by Marble 1962, 1970) through the perturbation of a uniform stream is applied to the flow over a wavy wall (Marble 1962). As in gas dynamics (Sears 1954), the gas-particle flow equivalent of the Prandtl–Glauert equation is obtained in the special case when the situation is steady in the moving frame of reference. This alternate derivation essentially connects the acoustics and aerodynamic concepts of small perturbation theory in gas-particle flows.

In the process of obtaining the small perturbation theory, (a) the role of entropy production, if any and, (b) the possibility of expressing the perturbation velocities of both gas and particle phases in terms of the gradients of their respective velocity potentials, both arise. The latter was apparently overlooked in Chu and Parlange (1962).

3 Small Perturbation Equations in a Stationary Frame

Consider perturbations about a gas-particle medium at rest and in thermodynamic equilibrium. Denote the stationary coordinate system by $(\widetilde{x}_j, \widetilde{t})$; the undisturbed gas and particle phase quantities are at rest and are uniform, with subscript 0. The small perturbation expansion, about the stationary constant state,

$$p = p_0 + p', \quad \rho = \rho_0 + \rho', \quad T = T_0 + T', \quad u_i = 0 + u'_i$$
$$\rho_P = \rho_{P0} + \rho'_P, \quad T_P = T_0 + T'_P, \quad u_{Pi} = 0 + u'_{Pi} \qquad (3.1),$$

then yields, for the gas phase conservation equations for continuity, momentum, and state from (2.5), for energy from (2.6) (after dropping the primes for convenience), and in stationary coordinates,

$$\frac{\partial \rho}{\partial \widetilde{t}} + \rho_0 \frac{\partial u_j}{\partial x_j} = 0$$

$$\rho_0 \frac{\partial u_i}{\partial \widetilde{t}} = -\frac{\partial P}{\partial \widetilde{x}_i} + \frac{\rho_{P0}}{\tau_{V0}}(u_{Pi} - u_i)$$

$$\rho_0 c_P \frac{\partial T}{\partial \widetilde{t}} = \frac{\partial P}{\partial \widetilde{t}} + \frac{\rho_{P0}}{\tau_{T0}} c_S (T_P - T) \qquad (3.2).$$

$$\frac{p}{p_0} = \frac{\rho}{\rho_0} + \frac{T}{T_0}$$

The heat capacity of the gas is taken as a constant averaged quantity. Substituting the perturbation expansion into the particle conservation equations

(2.2), (2.5), and (2.6) leads to the following first-order perturbation equations in stationary coordinates:

$$\frac{\partial \rho_P}{\partial \tilde{t}} + \rho_{P0} \frac{\partial u_{Pj}}{\partial \tilde{x}_j} = 0$$

$$\rho_{P0} \frac{\partial u_{Pi}}{\partial \tilde{t}} = \frac{\rho_{P0}}{\tau_{V0}} (u_{Pi} - u_i) \qquad (3.3).$$

$$\rho_{P0} c_S \frac{\partial T_P}{\partial \tilde{t}} = -\frac{\rho_{P0}}{\tau_{T0}} c_S (T_P - T)$$

3.1 The Role of Entropy

It is particularly important to discuss the mechanism of net entropy production, if any, in terms of small perturbation theory. There are in general two sources for the net production of entropy when the particle and gas phases are not in momentum and thermal equilibrium. One is the work done on the gas by velocity slip between the two phases and is the second term on the right side of the gaseous entropy equation from (2.17):

$$(u_{Pi} - u_i) F_{Pi} / T > 0,$$

which is a positive definite quantity. This dissipative work enters in the second order only: it is absent from first-order small perturbation theory.

The second entropy production mechanism is due to the inter-phase heat transfer Q_P. It occurs on the right sides of both the gaseous and particle phase entropy equations in (2.17). Normally heat transfer to the gas is received by the gas at its local temperature T and the local increase or decrease of gaseous entropy per unit volume is Q_P / T. Similarly, the local increase or decrease of particle phase entropy per unit volume is $-Q_P / T_P$, and the transfer of heat to and from the particle phase occurs at its temperature, T_P. To fix ideas, suppose that locally, $T > T_P$, and heat is locally transferred from the gas to the particle phase. In other words, the transfer of heat from the gaseous phase takes place at T and is received by the particle phase at a lower temperature, T_P. In this process, net entropy is produced. But for first-order small perturbation theory, the heat transfer takes place at the free-stream temperature, T_0, and thus the mechanism for net entropy production is absent from the first-order theory.

3.2 Velocity Potentials

We proceed again from the situation when the gas and particle phases are initially at rest and in thermodynamic equilibrium – that is, from the acoustical situation. It is now convenient to use vector rather than index notation. The momentum equation for the gas phase, in terms of perturbation variables from (3.2), is:

$$\frac{\partial \vec{u}}{\partial \tilde{t}} = -\frac{1}{\rho_0}\widetilde{\nabla}p + \frac{\kappa_0}{\tau_{V0}}(\vec{u}_P - \vec{u}) \tag{3.4}$$

where the gradient operator $\widetilde{\nabla}$ is in the stationary coordinate system, and we denoted by $\kappa_0 = \rho_0/\rho_{P0}$ the initial undisturbed mass ratio of the particle phase to the gas phase in a unit volume of mixture. The momentum equation for the particle phase from (3.3) in vector form is:

$$\frac{\partial \vec{u}_P}{\partial \tilde{t}} = -\frac{1}{\tau_{V0}}(\vec{u}_P - \vec{u}) \tag{3.5}$$

Denote the gas-phase vorticity vector by $\vec{\Omega} = \vec{\nabla} \times \vec{u}$. Similarly, we introduce the analogous vorticity vector of the particle phase, $\vec{\Omega}_P = \vec{\nabla} \times \vec{u}_P$. If we perform the rotational vector operation on both sides of (3.4) and (3.5), we obtain, respectively:

$$\frac{\partial \vec{\Omega}}{\partial \tilde{t}} = \frac{\kappa_0}{\tau_{V0}}(\vec{\Omega}_P - \vec{\Omega})$$

$$\frac{\partial \vec{\Omega}_P}{\partial \tilde{t}} = -\frac{1}{\tau_{V0}}(\vec{\Omega}_P - \vec{\Omega}) \tag{3.6}$$

The two vorticity equations combine to give $\vec{\Omega} + \kappa_0\vec{\Omega}_P = \vec{F} + \vec{F}_P$, where $\vec{F}, \vec{F}_P$ are the initial values and remain constant following the respective particle paths. If they are zero initially, then they remain so. Then the respective velocities of the two phases can be expressed as the gradient of their respective velocity potentials:

$$\vec{u} = \widetilde{\nabla}\phi$$
$$\vec{u}_P = \widetilde{\nabla}\phi_P \tag{3.7}$$

This demonstrates an important theorem due to Marble (1962, 1970) for first-order small perturbation theory – that the first-order perturbation potentials ϕ, ϕ_P exist. This existence seems to have been overlooked in the pioneering work of Chu and Parlange (1962).

3.3 The Equation for Acoustic Propagation

From the conservation equations for both phases, (3.2), (3.3), and the definition of velocity potential (3.7), one obtains the equation for acoustic propagation in a stationary gas-particle medium (Marble 1962, 1970),

$$\left[\frac{\partial^2}{\partial \widetilde{t}^2}\left(\frac{\partial^2}{\partial \widetilde{t}^2} - a_{F0}^2\widetilde{\nabla}^2\right) + \frac{1+\kappa_0}{\tau_{V0}}\frac{\partial}{\partial \widetilde{t}}\left(\frac{\partial^2}{\partial \widetilde{t}^2} - a_{V0}^2\widetilde{\nabla}^2\right) + \frac{1+\gamma\kappa_0 c_S/c_P}{\tau_{T0}}\right.$$
$$\left.\frac{\partial}{\partial \widetilde{t}}\left(\frac{\partial^2}{\partial \widetilde{t}^2} - a_{T0}^2\widetilde{\nabla}^2\right) + \frac{(1+\kappa_0)(1+\gamma\kappa_0 c_S/c_P)}{\tau_{V0}\tau_{T0}}\left(\frac{\partial^2}{\partial \widetilde{t}^2} - a_{E0}^2\widetilde{\nabla}^2\right)\right]\phi = 0 \quad (3.8).$$

Chu and Parlange (1962) obtained the same wave operator, although not in terms of the potential function. The Laplacian operator in stationary coordinates is $\widetilde{\nabla}^2$. Equation (3.8) appears in a higher-order form in the relaxation wave equation studied by Whitham (1959, 1974).

At least three acoustic propagation speeds occur in (3.8). The obvious one is the frozen speed of sound, a_{F0}, the propagation speed in the undisturbed state of the gas as if no particles are present. It is given by:

$$a_{F0}^2 = \gamma\frac{p_0}{\rho_0} \qquad (3.9).$$

It is associated with the highest-order wave operator in (3.8).

The lowest-order wave operator is associated with the equilibrium acoustic speed, a_{E0}, defined as if the gas and particle phases are in complete momentum and thermodynamic equilibrium, in which case the density is ρ_m so that:

$$a_{E0}^2 = \overline{\gamma}\frac{p_0}{\rho_m} \qquad (3.10),$$

where the equilibrium heat capacity ratio is appropriately defined as for the equilibrium mixture as:

$$\overline{\gamma} = \frac{\rho_0 c_P + \rho_{P0}c_S}{\rho_0 c_V + \rho_{P0}c_S} = \frac{1+\kappa_0 c_S/c_P}{1+\kappa_0\gamma c_S/c_P}\gamma \qquad (3.11).$$

The equilibrium density is simply $\rho_m = \rho_0(1+\kappa_0)$, and the equilibrium pressure is just p_0 since the particle phase makes no contribution to the pressure. Then the equilibrium sound speed becomes:

$$a_{E0}^2 = \frac{1}{(1+\kappa_0)}\frac{1+\kappa_0 c_S/c_P}{1+\kappa_0\gamma c_S/c_P}a_0^2 \qquad (3.12).$$

The wave operator in (3.8) that is operated on by a first-order time derivative can, for lack of a better term, be called an intermediate wave operator. Although a single acoustical speed would suffice, Marble (1970) points out that the "intermediate acoustic speed" can be interpreted physically as the sum of two parts. One is in the case when $\tau_{V0}\to 0$, τ_{T0} is finite, in this case $u_i\to u_{Pi}$, and the effective mixture density is $\rho_m = (1 + \kappa_0)\rho_0$, so that the velocity-equilibrium acoustical speed, a_{VE0}, is:

$$a_{VE0}^2 = \gamma \frac{p_0}{\rho_m} = a_{F0}^2/(1 + \kappa_0) \tag{3.13}.$$

For the intermediate wave operator, in the case of only thermodynamic equilibrium – that is, $\tau_{T0} \to 0$, τ_{V0} finite – then $T_P \to T$, and the effective heat capacity ratio is $\bar{\gamma}$ already defined in (3.11). In this case, the thermodynamic-equilibrium acoustic wave speed, a_{TE}, is

$$a_{TE0}^2 = \bar{\gamma} \frac{p_0}{\rho_0} = a_{F0}^2 \frac{1 + \kappa_0 c_S/c_P}{1 + \gamma \kappa_0 c_S/c_P} \tag{3.14}.$$

Examining the wave speeds in (3.9) and (3.12)–(3.14) shows that the higher-order wave speeds are progressively greater than the lower-order ones. Thus Whitham's (1959) wave hierarchy stability condition is satisfied; the domain of dependence is characterized by the highest-order wave at the frozen acoustic speed, a_{F0}.

While it is physically instructive to identify the intermediate waves with the two physical mechanisms for partial equilibrium wave speeds, $a_{V0}(\tau_{V0} \to 0), a_{T0}(\tau_{T0} \to 0)$ in (3.8), they are of the same third order. In studying wave structures, it is convenient to combine the intermediate waves into one effective single wave with a single effective (though not necessarily "physical") wave speed. If one introduces the single effective intermediate acoustic speed, a_I, as:

$$a_{I0}^2 = a_{F0}^2 \frac{\tau_{T0} + (1 + \kappa_0 c_S/c_P)\tau_{V0}}{(1 + \kappa_0)\tau_{T0} + (1 + \gamma \kappa_0 c_S/c_P)\tau_{V0}} \tag{3.15},$$

then the wave hierarchy equation (3.8) becomes, in more compact form,

$$\left[\frac{\partial^2}{\partial \tilde{t}^2} \left(\frac{\partial^2}{\partial \tilde{t}^2} - a_{F0}^2 \widetilde{\nabla}^2 \right) + \left(\frac{1 + \kappa_0}{\tau_{V0}} + \frac{1 + \gamma \kappa_0 c_S/c_P}{\tau_{T0}} \right) \frac{\partial}{\partial \tilde{t}} \left(\frac{\partial^2}{\partial \tilde{t}^2} - a_{I0}^2 \widetilde{\nabla}^2 \right) \right.$$
$$\left. + \frac{(1 + \kappa_0)(1 + \gamma \kappa_0 c_S/c_P)}{\tau_{V0}\tau_{T0}} \left(\frac{\partial^2}{\partial \tilde{t}^2} - a_{E0}^2 \widetilde{\nabla}^2 \right) \right] \phi = 0 \tag{3.16}.$$

The effective intermediate acoustic speed so defined does not change Whitham's (1959) wave hierarchy stability condition since $a_{F0} > a_{I0} > a_{E0}$. The zone of action is still determined by the highest-order wave speed, a_{F0}. The perturbation pressure, p, follows from the momentum equation for the gas:

$$\rho_0 \frac{\partial \phi}{\partial \tilde{t}} = -p + \frac{\kappa_0}{\tau_{V0}} (\phi_P - \phi) \tag{3.17},$$

which is coupled to ϕ_P through the potential form of the particle-phase momentum equation:

$$\frac{\partial \phi_P}{\partial \tilde{t}} = -\frac{1}{\tau_{V0}} (\phi_P - \phi) \tag{3.18}.$$

Equations (3.16)–(3.18) play the same role here as Rayleigh's [1894] (1945) acoustic propagation equations in a clean gas: they describe flows in aerodynamics generated by small perturbations external to boundary layers and in the absence of shock waves. In the following sections, a linearized theory for flow over thin bodies is obtained.

4 Aerodynamic Interpretation of Acoustics in Gas-Particle Flows

The acoustic propagation theory of Section 3 is used as a starting point to develop descriptions of flows and wave structures over thin bodies in gas-particle flows. This involves the statement that, in the frame of reference fixed on an observer moving with the thin body, the flow is described by the acoustic propagation equations.

The thin obstacle is considered as a source of small disturbances, or acoustic disturbances, and the resulting flow field is built up by superposition of such small disturbances. The flow field in the absence of the obstacle is a steady (or unsteady) uniform parallel stream at velocity u_0 along the positive x-direction. The reference frame fixed on the obstacle is represented by (x, y, z, t). The Galilean transformation fixing the reference frame on the obstacle is related to the stationary frame $(\tilde{x}, \tilde{y}, \tilde{z}, \tilde{t})$ by:

$$x = \tilde{x} + u_0 \tilde{t}, \ y = \tilde{y}, \ z = \tilde{z}, \ t = \tilde{t} \tag{4.1},$$

and:

$$\frac{\partial}{\partial x_j} = \frac{\partial}{\partial \tilde{x}_j}, \quad \frac{\partial}{\partial \tilde{t}} = \frac{\partial}{\partial t} + u_0 \frac{\partial}{\partial x} \tag{4.2}.$$

The corresponding gradient, divergence, and rotation operators are invariant:

$$\tilde{\nabla} = \nabla, \ \tilde{\nabla} \times (\ ) = \nabla \times (\ ) \tag{4.3}.$$

In the moving reference frame, (3.16) becomes:

$$\left(\frac{1}{u_0}\frac{\partial}{\partial t}+\frac{\partial}{\partial x}\right)^2\Pi_{F0}^2+\left(\frac{1}{\lambda_{V0}}+\frac{1+\gamma\kappa_0 c_S/c_P}{\lambda_{T0}}\right)\left(\frac{1}{u_0}\frac{\partial}{\partial t}+\frac{\partial}{\partial x}\right)\left[\frac{\partial^2}{\partial x^2}\Pi_{F0}^2\right.$$

$$\left.+\left(\frac{1}{\lambda_{V0}}+\frac{1+\gamma\kappa_0 c_S/c_P}{\lambda_{T0}}\right)\frac{\partial}{\partial x}\Pi_{I0}^2+\frac{(1+\kappa_0)(1+\gamma\kappa_0 c_S/c_P)}{\lambda_{V0}\lambda_{T0}}\Pi_{E0}^2\right]\phi=0$$

(4.4),

where each of the shorthand, unsteady Miles (1959) wave operators for the three Mach waves in (4.4) is defined as:

$$\Pi_{K0}^2(\,)\equiv\left[(1-M_{K0}^2)\frac{\partial^2}{\partial x^2}+\nabla_{yx}^2-2\frac{M_{K0}}{a_{K0}}\frac{\partial^2}{\partial x\partial t}-\frac{1}{a_{K0}^2}\frac{\partial^2}{\partial t^2}\right](\,)$$

(4.5),

where the subscripts $K = F, I,$ or E indicate the three wave operators. The sectional Laplacian,

$$\nabla_{yz}^2=\frac{\partial^2}{\partial y^2}+\frac{\partial^2}{\partial z^2}$$

(4.6),

is the Laplacian operator in the y,z-plane. The frozen, intermediate, and equilibrium Mach numbers introduced in (4.4) are respectively defined as $M_{K0} = u_0/a_{K0}$, for $K = F, I, E$ are such that $M_{F0} < M_{I0} < M_{E0}$. In this case, for supersonic flow, the Mach wave system, like the acoustic wave system, satisfies Whitham's (1959) wave hierarchy stability criterion. The streamwise momentum and thermal equilibrium distances are introduced as $\lambda_{V0} = u_0\tau_{V0}$, $\lambda_{T0} = u_0\tau_{T0}$. The "zone of action" is determined by the highest-order frozen Mach wave.

The perturbation pressure from (3.17) and (3.18) transforms respectively to

$$\left(\frac{1}{u_0}\frac{\partial}{\partial t}+\frac{\partial}{\partial x}\right)\phi=-\frac{p}{\rho_0 u_0}+\frac{\kappa_0}{\lambda_{V0}}(\phi_P-\phi)$$

(4.7)

$$\left(\frac{1}{u_0}\frac{\partial}{\partial t}+\frac{\partial}{\partial x}\right)\phi_P=-\frac{1}{\lambda_{V0}}(\phi_P-\phi)$$

(4.8).

The fundamental equations for unsteady small perturbation theory in two-phase gas-particle flow are given by (4.4), (4.7), and (4.8). The form of the individual wave operators in (4.4) is familiar in unsteady single-phase gas dynamics (Miles 1959).

For an obstacle exerting harmonic motions at frequency ω, while in a uniform stream at constant free-stream velocity u_0, the perturbation potentials and the perturbation pressure are then written as the product of an oscillatory part and a steady spatially dependent part in the form $z(x_j, t) = e^{i\omega t} Z(x_j)$. The resulting differential equations are easily obtained with the time derivative replaced by $i\omega$.

We reiterate that the wave system is hyperbolic when $M_{F0} > 1$, in which case the "zone of action" is determined by the highest-order wave and is the frozen Mach wave conoid of semi-vertex angle given by $\tan^{-1}\left(1/\sqrt{M_{F0}^2 - 1}\right)$.

5 Steady Small Perturbation Theory

Steady small perturbation theory in a uniform flow can be obtained directly via perturbation about a uniform stream, which leads to the Prandtl–Glauert–Ackeret type equation obtained by Marble (1970). In this section, small perturbation theory is unified with acoustics in a gas-particle flow by specializing equations (4.4), (4.5), (4.7), and (4.8) for steady flow by dropping the time derivatives (Sears 1954). Thus, for steady flow,

$$\left[\frac{\partial^2}{\partial x^2} \Pi_{F0}^2 + \left(\frac{1}{\lambda_{V0}} + \frac{1 + \gamma\kappa_0 c_S/c_P}{\lambda_{T0}} \right) \frac{\partial}{\partial x} \Pi_{I0}^2 \right.$$

$$\left. + \frac{(1 + \kappa_0)(1 + \gamma\kappa_0 c_S/c_P)}{\lambda_{V0}\lambda_{T0}} \Pi_{E0}^2 \right] \phi = 0 \tag{5.1},$$

where the steady state wave operators in (5.1) assume the familiar Prandtl–Glauert–Ackeret form:

$$\Pi_{K0}^2 = \left(1 - M_{K0}^2\right) \frac{\partial^2}{\partial x^2} + \nabla_{yz}^2 \tag{5.2},$$

for $K = F, I, E$. The relations for the pressure perturbation in the gas, (4.7) and (4.8), become:

$$\frac{\partial \phi}{\partial x} = -\frac{p}{\rho_0 u_0} + \frac{\kappa_0}{\lambda_{V0}} (\phi_P - \phi) \tag{5.3}$$

$$\frac{\partial \phi_P}{\partial x} = -\frac{1}{\lambda_{V0}} (\phi_P - \phi) \tag{5.4}.$$

In the present context, this derivation essentially connects the acoustic and aerodynamic concepts of small perturbation theory in gas-particle flows. The system of equations (5.3)–(5.4) holds for both subsonic and supersonic free-stream Mach numbers. In detailed applications to Mach wave structures, the equivalent Ackeret (1925) problem of the aerodynamics of supersonic flow over a very thin sharp two-dimensional body will be studied in detail. This appears to be the simplest problem to be studied from which subsequent extensions and improvements can be made.

Equations (5.3) and (5.4) for the perturbation pressure need to be improved if one discusses a slender body of revolution, particularly the pressure in the vicinity of the body surface where the perturbation velocities are not of the same order of magnitude (Liepmann and Roshko 1957) because of geometrical, curvature effects. However, equations (5.3) and (5.4) are meant to be applied to planar systems.

6 Some Limiting Cases

Two equilibration processes enter into the present consideration. They are the simultaneous occurrence of the velocity equilibration time, τ_{V0}, and thermal equilibration time, τ_{T0}. If the particles obey Stokes's law, then $\tau_{T0}/\tau_{V0} = (3/2)\mathrm{Pr}c_S/c_P$. In general, for small solid particles in gases, this ratio is of order unity, and this will be taken as the case in our detailed considerations. However, it is of interest to note the consequences of the limiting values of the time-scale ratios τ_{T0}/τ_{V0}, which readily yield to interpretations of the relaxation wave equation for a single relaxation (first derived by Stokes [1851] in a different context).

In considering the limiting cases, we again proceed from acoustic propagation. The equation for small perturbations in a uniform flow follow from a Galilean transformation.

For slow-velocity equilibration in a thermal equilibrium gas-particle flow, $\tau_{T0}/\tau_{V0}\to\infty$, then (3.16) reduces to:

$$\left[\frac{\tau_{V0}}{1+\kappa_0}\frac{\partial}{\partial\tilde{t}}\left(\frac{\partial^2}{\partial\tilde{t}^2}-a_{F0}^2\widetilde{\nabla}^2\right)+\left(\frac{\partial^2}{\partial\tilde{t}^2}-a_{VE0}^2\widetilde{\nabla}^2\right)\right]\phi = 0 \qquad (6.1),$$

where the velocity equilibrium acoustic propagation speed, a_{VE0}, was identified earlier (3.13). In obtaining (6.1), time integration is performed once and the spatially dependent constant of integration is discarded, implying that no initial disturbance is present. In the opposite limiting case, $\tau_{T0}/\tau_{V0}\to0$, a similar (except for details) equation is obtained:

$$\left[\frac{\tau_{T0}}{1+\gamma\kappa_0 c_S/c_P}\frac{\partial}{\partial\tilde{t}}\left(\frac{\partial^2}{\partial\tilde{t}^2}-a_{F0}^2\widetilde{\nabla}^2\right)+\left(\frac{\partial^2}{\partial\tilde{t}^2}-a_{TE0}^2\widetilde{\nabla}^2\right)\right]\phi = 0 \qquad (6.2),$$

where the thermal equilibrium acoustic speed, a_{TE0}, was identified in (3.14). Equation (6.2) describes the thermal equilibration process in a dynamically frozen situation. The single relaxation time equation is also obtained when the momentum and thermal equilibration parameters are identical.

Equations (6.1) and (6.2) are identical in form to acoustic propagation in a reacting gas involving a single reaction or relaxation time. The equation in this form was first derived by Stokes (1851) in his consideration of the effect of radiation of heat on the propagation of sound. But the "Stokes relaxation acoustic equation" seems to have escaped the attention of modern investigators. Stokes (1851) also obtained the periodic solution, equivalent to the flow over a wavy wall in the Prandtl–Glauert–Ackeret context, which was worked out by Vincenti (1959). Equation (6.1), or (6.2), is extensively discussed by Chu (1957), Whitham (1959) – who discussed more general forms of the relaxation wave equation involving a single relaxation time – and Moore and Gibson (1960). Much has been learned about wave structures from this work. For steady supersonic flow over a thin disturbance, for instance, the highest-order Mach wave is the frozen Mach wave, along which signals damp exponentially while the bulk of the disturbance is carried by the equilibrium Mach wave of lower order, which has a diffusive structure. Experimental aspects are reported by Wegener and Cole (1962). In what follows, we show how a two-relaxation time (or length) wave structure behaves.

7 Two-Dimensional Steady Supersonic Flow

So far, we have discussed in rather general terms small perturbation theory in gas-particle flows. In this section, the two-dimensional form of equation (5.1) and its ancillary equations are applied to specific discussion of steady, supersonic flow past very thin and sharp obstacles, which is the corresponding Ackeret (1925) problem. In that work, it is well known that its simplicity lies in the fact that the pressure acting on an element of surface depends only on the local surface deflection. In this case, the aerodynamic forces on a two-dimensional, thin obstacle are easily obtained through a simple integration when the local slope is prescribed as a function of the streamwise distance.

The fact that this simplicity no longer exists in gas-particle flows is due to the particle-gas equilibration process, which depends on upstream history. In this situation, the aerodynamic forces then depend on the extent to which the equilibration processes take place over the finite surface length of the obstacle, denoted by C. The natural parameters that arise are the ratios of the streamwise characteristic length C of the obstacle to the equilibration distances: C/λ_{V0}, C/λ_{T0}. If C/λ_{V0}, $C/\lambda_{T0} \gg 1$, the equilibration processes are confined to the surface of the obstacle and adjust relatively quickly to the local environment, provided that the surface shape is "slowly varying." When C/λ_{V0}, $C/\lambda_{T0} \ll 1$, the disturbances induced by the obstacle make themselves felt through the equilibration processes far into the downstream regions of the wake. As far as the surface pressures are concerned, one need only consider the region bounded by the frozen Mach waves, which determine the zone of action emanating from the leading and trailing edges of the obstacle.

Chu and Parlange (1962) used the relaxation acoustic equation (3.8) to discuss the one-dimensional unsteady motion in which a piston is suddenly set into motion and is subsequently maintained in steady motion. This would correspond to the situation of a simple semi-infinite wedge in a steady, two-dimensional supersonic flow, where the streamwise distance plays the role of time. They showed that the disturbance, while it decays along the frozen wave front defined by the frozen sound speed, a_{F0}, is ultimately propagated along the wave front defined by the equilibrium sound speed . The pressure on the piston face for finite times is not given, however, and their consideration of the wave structure for "large" times does not yield a physically recognizable form.

The present discussion of small perturbations in supersonic flow takes an aerodynamic point of view, providing a complementary (and complete) depiction of the acoustical piston problem addressed by Chu and Parlange (1962).

7.1 Mach Wave Structure

The two-dimensional form of the steady flow equation is obtained from (5.1) as

$$\left[\frac{\partial^2}{\partial x^2} \left(\beta_{F0}^2 \frac{\partial^2}{\partial x^2} - \frac{\partial^2}{\partial y^2} \right) + \left(\frac{1}{\lambda_{V0}} + \frac{1 + \kappa_0 c_S / c_P}{\lambda_{T0}} \right) \right.$$
$$\left. \frac{\partial}{\partial x} \left(\beta_{I0}^2 \frac{\partial^2}{\partial x^2} - \frac{\partial^2}{\partial y^2} \right) + \left(\frac{1 + \kappa_0 c_S / c_P}{\lambda_{V0} \lambda_{T0}} \right) \left(\beta_{E0}^2 \frac{\partial^2}{\partial x^2} - \frac{\partial^2}{\partial y^2} \right) \right] \phi = 0 \qquad (7.1),$$

where

$$\beta_{F0}^2 = M_{F0}^2 - 1, \quad \beta_{I0}^2 = M_{I0}^2 - 1, \quad \beta_{E0}^2 = M_{E0}^2 - 1 \qquad (7.2).$$

The Mach wave inclinations $\beta_{F0} < \beta_{I0} < \beta_{E0}$ are positive for supersonic flow (see Figure 1).

We are reminded that the intermediate wave system was earlier combined into an effective single intermediate wave in view of the fact that the intermediate system, which can otherwise be split into physically identifiable waves, depending on the velocity or thermodynamic equilibrium, are of the same order – that is, they are of third order. The definition of an effective acoustic speed (3.15) and an intermediate effective Mach number enters the coefficients in (7.1) naturally different from those in Marble's (1970). The present form, (7.1), will prove more conducive to analyzing the wave structures.

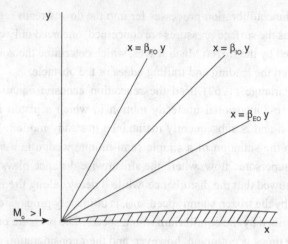

Figure 1 Mach wave locations in supersonic gas-particle flow. Schematic.

Zung (1967) found the exact solution for the flow over a wavy wall for the two-relaxation distance problem. The following analysis studies the Mach wave structure and is construed as an initial value problem in the streamwise distance with vanishing disturbance at $x = 0$ and conditions prescribed on $y = 0$.

We introduce the following dimensionless quantities:

$$x' = x/\lambda_{V0}, \ y' = y/\lambda_{V0}, \ \phi' = \phi/\lambda_{V0}u_0, \ \sigma = \lambda_{V0}(1 + \kappa_0 c_S/c_P)/\lambda_{T0} \quad (7.3),$$

and

$$u' = \frac{u}{u_0} = \frac{\partial \phi'}{\partial x'}, \ v' = \frac{v}{u_0} = \frac{\partial \phi'}{\partial y'} \quad (7.4).$$

Substituting (7.3) and (7.4) into (7.1), we have:

$$\left[\frac{\partial^2}{\partial x'^2} \left(\beta_{F0}^2 \frac{\partial^2}{\partial x'^2} - \frac{\partial^2}{\partial y'^2} \right) + (1+\sigma) \frac{\partial}{\partial x'} \left(\beta_{I0}^2 \frac{\partial^2}{\partial x'^2} - \frac{\partial^2}{\partial y'^2} \right) \right.$$
$$\left. + \sigma \left(\beta_{E0}^2 \frac{\partial^2}{\partial x'^2} - \frac{\partial^2}{\partial y'^2} \right) \right] \phi' = 0 \quad (7.5).$$

The complete initial and boundary conditions are:

$$\begin{aligned} x' = 0, \ y' > 0: \ & \phi' = \phi'_{x'} = \phi'_{x'x'} = \phi'_{x'x'x'} = 0 \\ x' > 0, \ y' = 0^+: \ & v' = \phi'_{y'} = f(x') \end{aligned} \quad (7.6).$$

Here, we consider the upper surface of the thin obstacle in $y' > 0$, which is independent of the lower surface, and the frozen Mach wave determines that the zone of action is inclined toward the downstream direction,

$x' - \beta_{F0}y' = $ constant; the local surface inclination is $f(x')$, for which $f(0) = 0$ for the sharp-nosed obstacle required by small perturbation theory. We deduce the wave structure via exact solutions first; approximate treatment of the wave structures follow in order to interpret what was deduced by the exact solutions.

Denote the Laplace transform (e.g., Churchill 1959) of $\phi'(x', y')$ by:

$$\Phi(s, y') = \int_0^\infty e^{-sx'} \phi'(x', y') dx' \tag{7.7}.$$

By using the vanishing initial disturbances in (7.5), the transformed equation (7.5) for the perturbation potential becomes:

$$\frac{d^2\Phi}{dy'^2} - S^2(s)\Phi = 0 \tag{7.8}$$

The solution of (7.8) is:

$$\Phi(s, y') = A(s)e^{-S_1(s)y'} + B(s)e^{-S_2(s)y'} \tag{7.9},$$

where the roots are:

$$\{S_1(s), S_2(s)\} = \pm \left[\frac{s^2 + (\beta_{I0}/\beta_{F0})^2(1 + \sigma)s + (\beta_{E0}/\beta_{F0})^2\sigma}{s^2 + (1 + \sigma)s + \sigma} \right]^{1/2} \beta_{F0}s \tag{7.10}.$$

The general solution (7.9), (7.10), like the relaxation wave equation (7.5), contains Mach waves leaning downstream and upstream. In order to select the simple-wave solution – that is, waves inclined in the downstream direction only – consider (7.10) at large s; we then have $S_1(s) \approx +\beta_{F0}s$, $S_2(s) \approx -\beta_{F0}s$. Thus the appropriate solution for Mach waves inclining toward the downstream direction is represented by the first term in (7.9), which is essentially the "outgoing waves." The second term in (7.9) is deleted since it represents the "incoming waves" from infinity where no disturbances are assumed to originate. The solution satisfying the transformed boundary condition $\Phi_{y'}(s, 0^+) = F(s)$ is then:

$$\Phi(s, y') = -\frac{F(s)}{S_1(s)} e^{-S_1(s)y'} \tag{7.11},$$

where

$$F(s) = \int_0^\infty e^{-sx'} f(x') dx' \tag{7.12}.$$

The solution in the physical plane requires the evaluation of the integral:

$$\phi'(x', y') = \frac{1}{2\pi i} \int_{L_1} \frac{-F(s)}{S_1(s)} e^{sx' - S_1(s)y'} ds \tag{7.13},$$

where L_1 is the Bromwich path parallel to the imaginary axis to the right of all singularities of $F(s)e^{-S_1(s)y'}/S_1(s)$. The behavior of the wave structure can be deduced directly in the (s, y') plane, following Lighthill and Whitham (1955) in their consideration of kinematic waves. It is convenient to discuss the wave structures in terms of the induced normal velocity,

$$V(s, y') = \Phi_y(s, y') = F(s)e^{-S_1(s)y'} \tag{7.14},$$

from which the relevant wave structures can be deduced. We examine, first, the behavior in the vicinity of the frozen Mach wave, which is the wave front that defines the zone of action since it is the highest-order wave (having the highest derivatives), and, second, the location of the wave along which the main disturbance will ultimately be carried out. The behavior near the frozen Mach wave is obtained by expanding $S_1(s)$ for large s:

$$-S_1(s) \approx -\beta_{F0}s - \beta_{F0}\frac{1+\sigma}{2}\left[\left(\frac{\beta_{I0}}{\beta_{F0}}\right)^2 - 1\right] + \vartheta\left(\frac{1}{s}\right) \tag{7.15},$$

which gives the following interpretation in the physical plane, keeping y' fixed:

$$v'(x', y') \approx f(x' - \beta_{F0}y')e^{-\frac{1+\sigma}{2}\left[\left(\frac{\beta_{I0}}{\beta_{F0}}\right)^2 - 1\right]\beta_{F0}y'}[1 + \vartheta(x' - \beta_{F0}y')] \tag{7.16},$$

describing the exponential decay of signals along the frozen Mach wave characterized by β_{F0} while its form remains unchanged, $f(x' - \beta_{F0}y')$. The frozen Mach wave, from (7.5), is of one order higher in the x'-derivative than the intermediate Mach wave characterized by β_{I0}. Examination of the damping factor in (7.16) shows that the exponential decay can be attributed to the presence of the intermediate Mach wave of one order lower. This situation is familiar in wave motions in a relaxing gas involving only a single relaxation time (Whitham 1959). The effect of the lowest-order wave, characterized by the equilibrium Mach wave, on the frozen Mach wave structure is negligible in the present problem, it being two orders lower.

We note here that the linearization process replaces the actual inclination of the frozen Mach wave, $\tan^{-1}(1/\beta_F)$, based on the local acoustic speed, by the constant value $\tan^{-1}(1/\beta_{F0})$ based on the free-stream acoustic speed. Thus the mechanism for wave steepening in which the most forward wave is overtaken by waves from the rear, which eventually leads to shock formation, is ruled out in this approximation.

Since the disturbance signals are damped along the frozen Mach wave, it is obvious that the bulk of the disturbance will be propagated along waves other than the most forward wave even though it actually determines the zone of action. In this case, we again follow standard procedures (Lighthill and Whitham 1955) to investigate the behavior represented by (7.14) for large x'. However, to study wave motions, y'/x' is kept fixed so as to look at the behavior of the various waves in their respective vicinities as $x' \to \infty$. We consider a simple physical situation in which the disturbances are generated by a semi-infinite wedge of half angle α. In this case, the dimensionless normal velocity becomes:

$$v'(x',0) = f(x') = \alpha H(x') \tag{7.17}$$

where $H(x')$ is the unit Heaviside function. Furthermore, let us consider the behavior of $v'_{x'}(x',y')$ instead of $v'(x',y')$, so that now the boundary condition becomes the Dirac delta-function (Lighthill 1958). Thus $v'_{x'}(x',y')$ can now be represented by the contour integral:

$$v'_{x'}(x',y') = \frac{\alpha}{2\pi i} \int_{L_1} e^{sx' - S_1(s)y'} ds \tag{7.18}$$

The behavior of $v'_{x'}(x',y')$ for $x' \to \infty$ can now be deduced by applying the saddle-point method. Let us write $m = y'/x'$ in which m is kept fixed. The dominant contribution to the value of the integral in (7.18) comes from the vicinity of the saddle point, through which the contour is chosen to pass, when the exponential factor (the integrand),

$$e^{x'(s - mS_1(s))} \tag{7.19}$$

is a maximum. The location of the saddle point s_0 is obtained from:

$$1 - m\frac{dS_1(s_0)}{ds} = 0 \tag{7.20}$$

from which one can solve, in principle, for s_0 in terms of m, $s_0 = s_0(m)$. However, in general this does not yield the asymptotic representation with the desired simple physical features. Instead, we first seek to determine the value of m if $v'_{x'}$ attains a maximum. In other words, we seek to locate the particular Mach wave in which the exponential factor of (7.19) is a maximum. This condition is simply:

$$\frac{d}{dm}[s_0 - mS_1(s_0)] = 0 \qquad (7.21),$$

which gives $ds_0/dm - S_1(s_0) - mdS_1(s_0)/dm = 0$. Since $dS_1(s_0)/dm = m^{-1}$ ds_0/dm from (7.20), $S_1(s_0) = 0$. Upon examining (7.10) for $S_1(s_0) = 0$, we find that the value of s_0 for this condition is $s_0 = 0$. Also, following from (7.10) and (7.20),

$$m = \frac{1}{dS_1(0)/ds} = \frac{1}{\beta_{E0}} \qquad (7.22),$$

is the location of the maximum disturbance $v'_{x'}(x', y')$, which is the equilibrium Mach wave $x' = \beta_{E0}y'$. In this case, we need only study the wave structure for large x' in the vicinity of the equilibrium Mach wave, in the vicinity of the relevant saddle point $s_0 = 0$. Since $S_1(s)$ is analytic at $s_0 = 0$, we expand $S_1(s)$ in a Taylor series about that point to yield:

$$-mS_1(s) = -a_1s + a_2s^2 - a_3s^3 + \ldots \qquad (7.23).$$

The integration contour is chosen to pass through $s = 0$ and the Bromwich path is simply along the imaginary axis for which:

$$s = is_I.$$

In this case, (7.18) becomes:

$$v'_x = \frac{\alpha}{2\pi}\int_{-\infty}^{+\infty} e^{ix'[(1-a_1)s_I + a_3s_I^3 + \ldots]}e^{-x'[a_2s_I^2 - a_4s_I^4 + \ldots]}ds_I \qquad (7.24),$$

where:

$$a_1 = \beta_{E0}y'/x'$$

$$a_2 = \frac{1+\sigma}{\sigma}\frac{\beta_{E0}}{2}\left[1 - \left(\frac{\beta_{I0}}{\beta_{E0}}\right)^2\right]\frac{y'}{x'} \qquad (7.25).$$

$$a_3 = \frac{\beta_{E0}}{2\sigma}\left\{\left[1 - \left(\frac{\beta_{F0}}{\beta_{E0}}\right)^2\right] - \frac{(1+\sigma)^2}{\sigma}\left[1 - \left(\frac{\beta_{I0}}{\beta_{E0}}\right)^2\right]\right\}\frac{y'}{x'}$$

Taking only the leading terms in (7.25) yields, for (7.24),

$$v'_{x'}(x',y') = \frac{\alpha}{\sqrt{2\pi}} \left[\frac{1}{\sqrt{2\pi}} \int_{-\infty}^{+\infty} e^{ix'(1-a_1)s_I} e^{-x'a_2 s_I^2} ds_I \right] \tag{7.26}.$$

The interpretation of the integral in the bracket in (7.26) is the familiar one in Fourier transforms (Sneddon 1951), simply:

$$v'_{x'}(x',y') = \frac{\alpha}{\sqrt{2\pi}} \left[\frac{1}{\sqrt{2a_2 x'}} e^{-\frac{(1-a_1)^2 x'^2}{4a_2 x'}} \right] = \frac{\alpha}{\sqrt{2\pi}} \left[\frac{e^{-\frac{(x'-\beta_{E0}y')^2}{2\frac{1+\sigma}{\sigma}\beta_{E0}\left[1-\left(\frac{\beta_{I0}}{\beta_{E0}}\right)^2\right]y'}}}{\sqrt{\frac{1+\sigma}{\sigma}\beta_{E0}\left[1-\left(\frac{\beta_{I0}}{\beta_{E0}}\right)^2\right]y'}} \right] \tag{7.27},$$

after inserting the definitions of (7.25). It is clear that the signal is a Gaussian centered about the equilibrium Mach wave characterized by β_{E0} and that the signal resembles a diffused signal: its effective diffusivity has contributions from the intermediate Mach wave, characterized by β_{I0} of one order higher, as is well known in the single-relaxation-distance situation. We can define an effective diffusivity as:

$$D_{12} = \frac{1+\sigma}{\sigma} \frac{\beta_{E0}}{2} \left[1 - \left(\frac{\beta_{I0}}{\beta_{E0}}\right)^2 \right] (\lambda_{V0} u_0) \tag{7.28},$$

since $\beta_{E0} > \beta_{I0}$, $D_{12} > 0$. In physical variables, the maximum amplitude of v_x decreases as:

$$\frac{\alpha u_0}{\sqrt{4\pi D_{12} y/u_0}} \tag{7.29},$$

as is well known in mass-diffusion or heat-diffusion problems (Carslaw and Jaeger 1948, 1959). The extent to which the diffusion-like region spreads out from the center of the equilibrium Mach wave, or diffusion layer width, is:

$$\sqrt{4\pi D_{12} y/u_0} \tag{7.30}.$$

Since the intermediary Mach wave extends to infinity as $x = \beta_{I0} y$, the diffusion layer ahead of the equilibrium Mach wave never reaches the intermediate Mach wave. The response of v' itself is obtained from (7.27) through a simple integration as:

$$v' = \frac{\alpha}{\sqrt{\pi}} \int_{-\infty}^{\chi} e^{-\bar{\chi}^2} d\bar{\chi} = \frac{\alpha}{2}(1 + erf\chi) \tag{7.31},$$

where *erf* χ is the error function of χ and from (7.27), $\chi = (x - \beta_{E0}y)/\sqrt{4D_{12}y/u_0}$. In the integration, we use the condition that disturbances "far ahead" have become negligible by decaying far upstream from the equilibrium Mach wave.

Examination of the effective diffusivity in (7.28) shows that the diffusion of the lowest-order waves, $x' = \beta_{E0}y'$, is attributed to the presence of waves one order higher, $x' = \beta_{I0}y'$, which are the intermediate Mach waves. This is also the familiar situation in wave motions with a single relaxation time (Whitham 1959). The effect of the highest-order waves on the wave structure in the vicinity of the equilibrium Mach wave is contained in the neglected higher-order terms in (7.24). They are examined later.

The behavior in the vicinity of the intermediate Mach wave is by inference a simultaneous damping with decay factor containing $[(\beta_{E0}/\beta_{I0})^2 - 1]$, owing to the one-order-lower equilibrium Mach wave, and diffusion, owing to the one-order-higher frozen Mach wave, with effective diffusivity containing the factor $[1 - (\beta_{F0}/\beta_{I0})^2]$. This can be demonstrated more simply through the approximations to the wave equation suggested by Whitham (1959). One obtains:

$$v'_{x'}(x',y') = \frac{\alpha}{\sqrt{2\pi}} \frac{e^{-\frac{(x'-\beta_{I0}y')^2}{2\beta_{I0}[1-(\beta_{F0}\beta_{I0})^2]y'/(1+\sigma)}}}{\sqrt{\beta_{I0}[1-(\beta_{F0}/\beta_{I0})^2]y'/(1+\sigma)}} e^{-\frac{\sigma}{1+\sigma}\frac{\beta_{E0}}{2}[(\beta_{E0}/\beta_{I0})^2-1]y'}$$

(7.32)

for the structure in the vicinity of the intermediate Mach wave, $x' = \beta_{I0}y'$.

7.2 Mach Wave Structure for Finite-Length Obstacle of Chord Length C

In considering disturbances generated by the presence of a thin, finite-length obstacle in the streamwise direction, some conclusions of a general nature can be drawn from the far-field behavior due to a flat-plate of finite length, C (C' is C normalized by λ_{V0}), at a small negative angle of attack, α. In this case, the boundary condition (7.17) for the normal velocity becomes the superposition of two Heaviside functions placed C apart,

$$v'(x',0;C') = f(x';C') = \alpha[H(x') - H(x' - C')] \tag{7.33},$$

and $v'_{x'}$ becomes the superposition of two Dirac delta functions, whose interior solution is represented by the contour integral as for (7.18):

$$v'_{x'}(x',y';C') = \frac{\alpha}{2\pi i}\int_{L_1}\left(1 - e^{-C's}\right)e^{sx'-S_1(s)y'}ds \tag{7.34}.$$

The far-field behavior is similarly estimated as:

$$v_x(x,y;C) \approx \frac{\alpha u_0}{\sqrt{4\pi D_{12}y/u_0}} \left[e^{-\frac{(x-\beta_{E0}y)^2}{4D_{12}y/u_0}} - e^{-\frac{(x-C-\beta_{E0}y)^2}{4D_{12}y/u_0}} \right] \tag{7.35},$$

which is the superposition of two Gaussians, one centered about the leading-edge equilibrium Mach wave and the other centered about the trailing-edge equilibrium Mach wave. The normal velocity is similarly obtained via integration and is:

$$v = \frac{\alpha u_0}{2} (erf\chi - erf\chi_C) \tag{7.36},$$

where

$$\chi_C = (x - C - \beta_{E0}y)/\sqrt{4D_{12}y/u_0} \tag{7.37}.$$

Thus (7.36) is the superposition of two smooth steps whose maximum slopes are centered about the leading-edge and trailing-edge equilibrium Mach waves, respectively. The controlling factor is the ratio of the chord length, C, to the diffusion layer width, $\sqrt{4\pi D_{12}y/u_0}$. First, suppose that $C/\sqrt{4\pi D_{12}y/u_0} >> 1$: the extent of the diffusion-like region is very close to the equilibrium Mach waves relative to the chord length over which the gas-particle interactions are very close to equilibrium. The diffusion-like regions are so thin that that one can write approximately:

$$v \approx \alpha u_0 [H(x' - \beta_{E0}y') - H(x' - C' - \beta_{E0}y')] \tag{7.38},$$

as if the disturbances were concentrated at the equilibrium Mach waves. More generally, one can simply write:

$$v \approx u_0 f(x' - \beta_{E0}y') \tag{7.39},$$

for $C/\sqrt{4\pi D_{12}y/u_0} >> 1$. In the opposite limiting case of very broad diffusion regions compared to the chord length, $C/\sqrt{4\pi D_{12}y/u_0} << 1$, the regions merge to form a single diffusion region and the far-field behavior then spreads out as $\sqrt{4\pi D_{12}y/u_0}$ centered about an equilibrium Mach wave emanating from the center of the plate. The amplitude of the disturbance decreases as:

$$v \approx \alpha u_0/\sqrt{4\pi D_{12}y/u_0} \tag{7.40}.$$

The far fields for the two limiting situations are depicted in Figure 2.

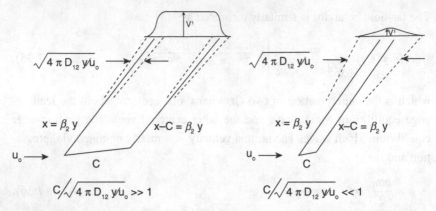

Figure 2 Far-field Mach wave structures for finite flat-plate of chord length C at negative angle of attack according to small-disturbance theory. Schematic.

7.3 The Far-Field Behavior for Finite-General Thin Body Shapes

We have shown the far-field behavior using the example of a flat-plate airfoil at an angle of attack, in which case the disturbances are propagated along the equilibrium Mach waves. Surrounding these Mach waves is a diffusion-like structure with an effective diffusivity D_{12}. The far field owing to other than the flat-plate airfoil at an angle of attack can be interpreted as due to a momentum source $\rho_0 u_0 v(x, 0)$. It's disturbances are ultimately diffused and propagated along equilibrium Mach lines. For a variable-surface, thin, sharp obstacle one can think of the far-field behavior as the superposition of sources that lie in the region $0 \le x \le C$ at $y = 0$, where the source strength per unit length is:

$$\rho_0 u_0 v(x, 0) = \rho_0 u_0 f(x) \tag{7.41}$$

in $0 \le x \le C$ and zero elsewhere. By analogy with heat conduction (Carslaw and Jaeger 1959), we may write, for the far field,

$$\rho_0 u_0 v(x, y) = \frac{\rho_0 u_0}{\sqrt{\pi}} \int\limits_{-\frac{x - \beta_{E0} y}{2\sqrt{D_{12} y / u_0}}}^{0} f\left(x - \beta_{E0} y + 2\xi \sqrt{D_{12} y / u_0}\right) e^{-\xi^2} d\xi$$

$$+ \frac{\rho_0 u_0}{\sqrt{\pi}} \int\limits_{0}^{\frac{C - x - \beta_{E0} y}{2\sqrt{D_{12} y / u_0}}} f\left(x - \beta_{E0} y + 2\xi \sqrt{D_{12} y / u_0}\right) e^{-\xi^2} d\xi \tag{7.42}$$

When $f(x) = constant$, as for a flat-plate at an angle of attack, relation (7.42) reverts to (7.36). We also conclude that, although the disturbances are no longer concentrated but are spread about the equilibrium Mach waves far from the obstacle, by analogy with the conduction of heat, the totality of the disturbance is nevertheless preserved.

7.4 Effects of the Frozen Mach Wave on the Equilibrium Mach Wave Structure

Thus far we have considered only the effect on the equilibrium Mach wave of the intermediate Mach wave of one order higher. The frozen Mach wave, which is of two orders higher than the equilibrium Mach wave in the fundamental equation (7.5), might at first exert a dispersion-like behavior on the structure about the equilibrium Mach wave. However, the presence of the intermediate Mach wave gives rise to the damping factor $e^{-x'a_2s_i^2}$ in (7.24) in the asymptotic behavior; this essentially submerges the oscillatory tendency contributed by the frozen Mach wave, embedded in the factor $e^{ix'a_3s_i^3}$ in (7.24). The exponential damping in (7.24) is unaffected by even the local oscillations. In fact, the result is monotonic. The formal aspects of this discussion are as follows.

We use (7.23) to expand the exponential of (7.24) in the form:

$$e^{-x'mS_1(s)} \cong e^{-x'a_1s+x'a_2s^2}\left(1 - x'a_3s^3 + \ldots\right) \tag{7.43}.$$

To consider the simple semi-infinite wedge of half-angle α, or the flat-plate at negative angle of attack at the same angle, the interpretation corresponding to the first term for $v'_{x'}$ has already been obtained. The second term has the formal interpretation (except for a multiplicative factor $x'a_3$) as the third $x'-$ derivative of the already known first term in the $x', y'-$ plane where the initial conditions in x' vanish. This is similar for other higher-order terms. Therefore:

$$v'_{x'}(x',y') \cong \frac{\alpha\lambda_{V0}}{\sqrt{4\pi D_{12}y/u_0}}$$

$$e^{-\chi^2}\left[1 - \frac{1+\sigma}{\sigma}\frac{\lambda_{V0}}{\sqrt{D_{12}y/u_0}}\left(\frac{D_{02}}{D_{12}} - 1\right)(3\chi - \chi^3) + \vartheta(1/y/\lambda_{V0})\right] \tag{7.44},$$

where

$$D_{02} = \frac{\lambda_{V0}u_0\beta_{E0}}{1+\sigma}\frac{\beta_{E0}}{2}\left[1 - \left(\frac{\beta_{F0}}{\beta_{E0}}\right)^2\right] \tag{7.45}$$

is an "effective diffusion coefficient" depicting the effect of the frozen Mach wave on the equilibrium Mach wave. The second term in (7.44) gives rise to the skewness in the Gaussian resulting in:

$$v' \cong \frac{a}{2}(1 + erf\chi) - \frac{a}{2}\frac{1+\sigma}{\sigma}\frac{\lambda_{V0}}{\sqrt{\pi D_{12}y/u_0}}\left(\frac{D_{02}}{D_{12}} - 1\right)(2\chi^2 - 1)e^{-\chi^2}$$

$$+ \vartheta(1/y/\lambda_{V0}) \qquad\qquad (7.46),$$

which shows the approach to the ultimate form of equilibrium wave structure indicated by the first term in (7.46). The effect of the frozen Mach wave is in the second term, which is diffusive in nature.

The present study thus far is based on the full relaxation wave equation for two relaxation lengths, (7.5). A more illuminating approach is one that makes direct approximations to the wave equation (7.5) itself (Whitham 1959), as well as addressing the perturbation problem for small particle loading ($\kappa_0 \ll 1$), for which the otherwise basic state of $\kappa_0 = 0$ is singular. These are addressed in the subsequent sections.

8 Approximate Consideration Based on a "Rarefied" Particle Cloud

A dilute particle phase in this context is one in which the particle phase density, $\rho_{P0} = n_{P0}m_P$, where n_{P0}, the number density of the particles per unit volume of the mixture, and m_P, the mass of a single particle, are small relative to the gas density, ρ_0. It is always assumed that there are enough particles in a unit volume to render valid the continuum description of the particle phase. The ratios of relaxation times, τ_{V0}/τ_{T0}, and that of the heat capacities, c_S/c_P (again, the heat capacities are taken as individually averaged-constants, even before linearization), are considered here to be of order unity as would be expected in the actual situation.

In this context, the inclinations of the frozen, intermediate, and equilibrium Mach waves are not significantly different from one another,

$$\varepsilon_1 \equiv \frac{\beta_{I0}}{\beta_{F0}} - 1 \cong \kappa_0 \frac{1 + \beta_{F0}^2}{2\beta_{F0}}\frac{1 + (\gamma - 1)(c_S/c_P)(\lambda_{V0}/\lambda_{T0})}{1 + (\lambda_{V0}/\lambda_{T0})} + \vartheta(\kappa_0^2) \qquad (8.1)$$

$$\varepsilon_2 \equiv \frac{\beta_{E0}}{\beta_{F0}} - 1 \cong \kappa_0 \frac{1 + \beta_{F0}^2}{2\beta_{F0}}[1 + (\gamma - 1)(c_S/c_P) + \vartheta(\kappa_0^2) \qquad (8.2),$$

where we use the previous definitions of acoustic speeds, from which expansions for $\kappa_0 \ll 1$ are performed. For the range of free-stream Mach numbers M_{F0}

consistent with small perturbation theory, the factor $\left(1 + \beta_{F0}^2\right)/2\beta_{F0} \approx \vartheta(1)$, and hence $\varepsilon_1, \varepsilon_2 \approx \vartheta(\kappa_0), \quad \varepsilon_1/\varepsilon_2 \approx \vartheta(1)$. The consequence of this situation is that the Mach waves emanating from the same point, bounded by the frozen and equilibrium Mach waves, cluster very closely together, anticipating that changes across the Mach waves are steeper than changes along these waves (a "boundary layer" in the making).

One is then tempted to apply higher-order expansions for the dependent variables with κ_0 as a perturbation parameter; here $\kappa_0 = 0$ provides the zeroth-order approximation, beginning with the basic Ackeret problem:

$$\beta_{F0}^2 \phi_{xx} - \phi_{yy} = 0 \tag{8.3}.$$

On the other hand, one recognizes that in this approximation, the mutual effects of the waves upon each other are lost. Thus, this type of perturbation is then obviously a singular one, and it does not provide a smooth transition from the frozen to the equilibrium state.

In deriving a simplified but uniformly valid differential equation from the full relaxation wave equation (7.5) for the dilute particle phase, we begin first from a physically intuitive approach pioneered by Whitham (1959). We show subsequently that by properly applying the "boundary layer" concept in a transformed coordinate system, the resulting zeroth-order differential equation is the same as that obtained from the physical derivation. However, the formal Poincaré–Lighthill-like (Lighthill 1949) perturbation scheme enables one to demonstrate how systematic higher corrections can be obtained.

Within the limitations of the linearized theory, it is required that the free-stream Mach number be sufficiently larger than unity and the disturbance sufficiently small, and that the flow is everywhere supersonic. In the two-dimensional case, and in the three-dimensional case away from the tip-frozen Mach conoids, the upper surfaces of a thin obstacle are then independent of the lower surface. To fix ideas, consider the upper surface of a thin airfoil in a stream of infinite extent. The disturbances produced by the thin airfoil are carried along Mach waves inclined toward the downstream direction — that is, waves of the $x = \beta_{i0}y$ family. Assuming that there are no disturbances in the stream at infinity, we need only focus attention on waves of a single family in considering the upper surface. In the absence of particles, the single function $\phi(x - \beta_{F0}y)$ is retained from the general solution for the upper surface, while $\phi(x + \beta_{F0}y)$ is used for the lower surface (known as the simple-wave approximation). This suggests that for the relaxation wave problem, a similar isolation of Mach waves of a single family can be made and applied directly to the differential equation (7.5) itself.

8.1 A Physical Derivation

We essentially follow the spirit of Whitham's (1959) approximate treatment of wave motions for the single relaxation process. For convenience, we rewrite (7.5) into a form that exhibits the wave operators corresponding to the two families of Mach waves, $x = \pm\beta_{i0}y$, where the subscripts $i = F, I, E$ indicate frozen, intermediate, and equilibrium waves, respectively:

$$\left[\frac{\partial^2}{\partial x'^2}\left(\beta_{F0}\frac{\partial}{\partial x'} - \frac{\partial}{\partial y'}\right)\left(\beta_{F0}\frac{\partial}{\partial x'} + \frac{\partial}{\partial y'}\right) + (1 + \sigma)\frac{\partial}{\partial x'}\left(\beta_{I0}\frac{\partial}{\partial x'} - \frac{\partial}{\partial y'}\right)\right.$$
$$\left.\left(\beta_{I0}\frac{\partial}{\partial x'} + \frac{\partial}{\partial y'}\right) + \sigma\left(\beta_{E0}\frac{\partial}{\partial x'} - \frac{\partial}{\partial y'}\right)\left(\beta_{E0}\frac{\partial}{\partial x'} + \frac{\partial}{\partial y'}\right)\right]\phi' = 0 \qquad (8.4)$$

Consider now the upper surface $y' > 0$. The relevant wave operators to be retained are:

$$\beta_{i0}\frac{\partial}{\partial x'} + \frac{\partial}{\partial y'},$$

which describe waves inclined in the downstream direction. For this family of Mach waves, the "simple-wave approximation" amounts to evaluating the wave derivatives occurring elsewhere by:

$$\beta_{i0}\frac{\partial}{\partial x'} \approx -\frac{\partial}{\partial y'}.$$

In addition, for the dilute particle phase:

$$\kappa_0 \ll 1, \quad (\beta_{I0}/\beta_{F0} - 1) \ll 1, \quad (\beta_{E0}/\beta_{F0} - 1) \ll 1.$$

Thus the derivatives in the "other" wave operators $\beta_{i0}\partial/\partial x' - \partial/\partial y'$ are approximated by:

$$-\frac{\partial}{\partial y'} \approx \beta_{F0}\frac{\partial}{\partial x'} \approx \beta_{I0}\frac{\partial}{\partial x'} \approx \beta_{E0}\frac{\partial}{\partial x'} \qquad (8.5).$$

Applying (8.5) to (8.4), after integrating with respect to x' once and discarding the integration constant (function of y'), yields the simple-wave approximation to the double-relaxation-length wave equation:

$$\left[\frac{\partial^2}{\partial x'^2}\left(\beta_{F0}\frac{\partial}{\partial x'} + \frac{\partial}{\partial y'}\right) + (1 + \sigma)\frac{\partial}{\partial x'}\left(\beta_{I0}\frac{\partial}{\partial x'} + \frac{\partial}{\partial y'}\right) + \sigma\left(\beta_{E0}\frac{\partial}{\partial x'} + \frac{\partial}{\partial y'}\right)\right]\phi' = 0$$
$$(8.6),$$

which describes Mach waves inclined in the downstream direction, with the simple-wave approximation applied to the original differential equation (8.4). This would have been done upon the much simpler Prandtl–Glauert–Ackeret equation for supersonic flow resulting, in the absence of the particle phase:

$$\left(\beta_{F0}\frac{\partial}{\partial x'} + \frac{\partial}{\partial y'}\right)\phi' = 0.$$

8.2 Derivation via Strained Coordinates

Implicit in the physical derivation in the previous section is that changes of flow variables across the downstream-running Mach waves, in $x' - \beta_{F0}y'$, are much more rapid than changes along the waves in, say, y'. This is the spirit of the "boundary layer" concept Moore and Gibson (1960) suggested in their consideration of the single-relaxation-process Mach waves (more generally, in the spirit of Carrier 1953, 1954).

The "slowly varying" changes along the Mach waves are demonstrated by the exponential decay factor from the exact consideration of (7.15):

$$e^{-\frac{1+\sigma}{2}[(\beta_{I0}/\beta_{F0})^2 - 1]\beta_{F0}y'},$$

which, for $\kappa_0 \ll 1$, is:

$$e^{-(1+\sigma)\varepsilon_1\beta_{F0}y'},$$

where $\varepsilon_1 \approx \vartheta(\kappa_0)$ from (8.1). The "boundary layer" across the closely clustered Mach waves is more readily demonstrated after performing a transformation of the full small-disturbance relaxation wave equation (7.5) from the physical plane (x',y') to strained dependent variables across and along the waves, respectively:

$$\beta_{F0}^2\xi = x' - \beta_{F0}y', \; \beta_{F0}\eta = y', \tag{8.7}.$$

With the transformation rules from (8.7), we have:

$$\frac{\partial}{\partial x'} = \frac{1}{\beta_{F0}^2}\frac{\partial}{\partial \xi}, \; \frac{\partial}{\partial y'} = \frac{1}{\beta_{F0}}\left(\frac{\partial}{\partial \eta} - \frac{\partial}{\partial \xi}\right) \tag{8.8}.$$

Applying the coordinate transformation (8.8) to the wave equation (7.5), we obtain:

$$\left[\left(\frac{\partial^2}{\partial\xi^2} + (1+\sigma)\beta_{F0}^2\frac{\partial}{\partial\xi} + \sigma\beta_{F0}^4\right)\left(2\frac{\partial^2}{\partial\xi\partial\eta} - \frac{\partial^2}{\partial\eta^2}\right)\right.$$

$$+\left((1+\sigma)\beta_{F0}^2\frac{\partial}{\partial\xi} + \sigma\beta_{F0}^4\right)\left(\varepsilon_1 2\left(1+\frac{\varepsilon_2}{\varepsilon_1}\right)\right)$$

$$\left.+\varepsilon_1^2\left(1+\frac{\varepsilon_2^2}{\varepsilon_1^2}\right)\frac{\partial^2}{\partial\xi^2}\right]\phi' = 0 \tag{8.9},$$

where, we are reminded from (8.1) and (8.2), that $\varepsilon_1 = \vartheta(\kappa_0) \ll 1$, $\varepsilon_1/\varepsilon_2 = \vartheta(1)$. It is clear that using ε_1 identically to zero as a starting point is singular, if we let:

$$\bar{\bar{\xi}} = \varepsilon_1^a\xi, \quad \bar{\eta} = \varepsilon_1^b\eta \tag{8.10}.$$

In order that the coordinates be strained to render the differential equations uniformly valid (Carrier 1953, 1954), it is found that: $a = 0, \ b = 1$

$$\frac{\partial}{\partial\bar{\bar{\xi}}}\left[\frac{\partial^3}{\partial\bar{\eta}\partial\bar{\bar{\xi}}^2} + (1+\sigma)\beta_{F0}^2\frac{\partial}{\partial\bar{\bar{\xi}}}\left(\frac{\partial}{\partial\bar{\bar{\xi}}} + \frac{\partial}{\partial\bar{\eta}}\right) + \sigma\beta_{F0}^4\left(\frac{\varepsilon_2}{\varepsilon_1}\frac{\partial}{\partial\bar{\bar{\xi}}} + \frac{\partial}{\partial\bar{\eta}}\right)\right]\phi' =$$

$$\frac{\varepsilon_1}{2}\left[\frac{\partial^4}{\partial\bar{\eta}^2\partial\bar{\bar{\xi}}^2} + (1+\sigma)\beta_{F0}^4\frac{\partial}{\partial\bar{\bar{\xi}}}\left(\frac{\partial^2}{\partial\bar{\eta}^2} - \frac{\partial^2}{\partial\bar{\bar{\xi}}^2}\right) + \sigma\beta_{F0}^4\left(\frac{\partial^2}{\partial\bar{\eta}^2} - \frac{\varepsilon_2^2}{\varepsilon_1^2}\frac{\partial^2}{\partial\bar{\bar{\xi}}^2}\right)\right]\phi'$$

$$\tag{8.11}.$$

A systematic approximation for ϕ' can thus be made by expansion in ascending powers of the small parameter ε_1:

$$\phi' = \phi_0 + \varepsilon_1\phi_1 + \varepsilon_1^2\phi_2 + \ldots \tag{8.12}.$$

Substituting (8.12) into (8.11) gives, after one integration in $\bar{\bar{\xi}}$ and discarding an integration constant (function of $\bar{\eta}$) for outgoing wave motion, the zeroth order wave equation:

$$\left[\frac{\partial^3}{\partial\bar{\eta}\partial\bar{\bar{\xi}}^2} + (1+\sigma)\beta_{F0}^2\frac{\partial}{\partial\bar{\bar{\xi}}}\left(\frac{\partial}{\partial\bar{\bar{\xi}}} + \frac{\partial}{\partial\bar{\eta}}\right) + \sigma\beta_{F0}^4\left(\frac{\varepsilon_2}{\varepsilon_1}\frac{\partial}{\partial\bar{\bar{\xi}}} + \frac{\partial}{\partial\bar{\eta}}\right)\right]\phi_0 = 0$$

$$\tag{8.13}.$$

If we transform (8.13) back to the physical coordinates (x', y'), (8.6) is recovered. This essentially demonstrates the equivalence of the more formal procedure with the earlier physically intuitive derivation, which is a simple-wave approximation. The more formal scheme enables one to obtain higher-order corrections in ε_1, if desired, from the inhomogeneous differential equations:

$$\frac{\partial}{\partial\bar{\xi}}\left[\frac{\partial^3}{\partial\bar{\eta}\partial\bar{\xi}^2} + (1+\sigma)\beta_{F0}^2\frac{\partial}{\partial\bar{\xi}}\left(\frac{\partial}{\partial\bar{\xi}} + \frac{\partial}{\partial\bar{\eta}}\right) + \sigma\beta_{F0}^4\left(\frac{\varepsilon_2}{\varepsilon_1}\frac{\partial}{\partial\bar{\xi}} + \frac{\partial}{\partial\bar{\eta}}\right)\right]\phi_n =$$

$$\frac{1}{2}\left[\frac{\partial^4}{\partial\bar{\eta}^2\partial\bar{\xi}^2} + (1+\sigma)\beta_{F0}^4\frac{\partial}{\partial\bar{\xi}}\left(\frac{\partial^2}{\partial\bar{\eta}^2} - \frac{\partial^2}{\partial\bar{\xi}^2}\right) + \sigma\beta_{F0}^4\left(\frac{\partial^2}{\partial\bar{\eta}^2} - \frac{\varepsilon_2^2}{\varepsilon_1^2}\frac{\partial^2}{\partial\bar{\xi}^2}\right)\right]\phi_{n-1}$$

$$(8.14),$$

where $n = 1,\ 2,\ \ldots$ indicates the order of approximation. The appropriate condition at $\bar{\eta} = 0$ – that is, at $y' = 0$ – is satisfied by the zeroth-order solution ϕ_0. All higher-order corrections then satisfy homogeneous conditions at $\bar{\eta} = 0$. In what follows, we only consider the zeroth-order solution from equation (8.6).

8.3 Mach Wave Structure from Simplified Wave Equation

We proceed to consider the Mach wave structure as obtained from the simplified wave equation. The subscript 0 is dropped, as well as the prime, indicating dimensionless potential function. It is understood that the zeroth-order problem is considered. For convenience, (8.6) is repeated here:

$$\left[\frac{\partial^2}{\partial x'^2}\left(\beta_{F0}\frac{\partial}{\partial x'} + \frac{\partial}{\partial y'}\right) + (1+\sigma)\frac{\partial}{\partial x'}\left(\beta_{I0}\frac{\partial}{\partial x'} + \frac{\partial}{\partial y'}\right) + \sigma\left(\beta_{E0}\frac{\partial}{\partial x'} + \frac{\partial}{\partial y'}\right)\right]\phi = 0$$

$$(8.15).$$

The initial and boundary conditions are:

$$x' = 0,\ y' > 0:\quad \phi = \phi_{x'} = \phi_{xx'} = 0$$
$$x' > 0,\ y' = 0^+:\quad v' = \phi_{y'} = f(x')$$

$$(8.16).$$

As the boundary condition shows, we consider the upper surface, $y' > 0$, in a free-stream u_0 in the positive x' direction. The Laplace transformation,

$$\Phi(s,y') = \int_0^\infty e^{-sx'}\phi(x',y')dx'$$

$$(8.17),$$

is applied to (8.15). The transformed equation, after applying the initial conditions in x', is now a first-order ordinary differential equation in y':

$$\frac{d\Phi}{dy'} + S(s)\Phi = 0$$

$$(8.18),$$

where:

$$S(s) = \frac{s^2 + (\beta_{I0}/\beta_{F0})(1+\sigma)s + (\beta_{E0}/\beta_{F0})\sigma}{s^2 + (1+\sigma)s + \sigma}\beta_{F0}s$$

$$(8.19).$$

The boundary condition owing to the thin obstacle in the Laplace transformed variable is:

$$\Phi_{y'}(s, 0^+) = F(s) = \int_0^\infty e^{-sx'} f(x') dx' \tag{8.20}$$

Hence the solution is:

$$\Phi(s, y') = -\frac{F(s)}{S(s)} e^{-S(s)y'} \tag{8.21}$$

In physical variables, (8.21) is interpreted as:

$$\phi(x', y') = \frac{1}{2\pi i} \int_{L_1} -\frac{F(s)}{S(s)} e^{sx' - S(s)y'} ds \tag{8.22}$$

where, as in (7.13), L_1 is the Bromwich path. It is again convenient to study the wave structure in terms of the normal velocity:

$$V(s, y') = \Phi_{y'}(s, y') = F(s) e^{-S(s)y'} \tag{8.23}$$

Prior to deducing the asymptotic behavior of the wave structure from the simplified differential equation consideration, we obtain the operational solution for $v'(x', y')$ from (8.23). It can be shown that $S(s)$, as defined in (8.19), can be written in the form through the use of partial fractions:

$$(1 + \sigma)\varepsilon_1 \beta_{F0} + s\beta_{F0} - \Lambda_1 \frac{\beta_{F0}\varepsilon_1}{s + \sigma} - \Lambda_2 \frac{\beta_{F0}\varepsilon_1}{s + 1} \tag{8.24}$$

where

$$\Lambda_1 = \frac{\sigma^2}{1 - \sigma} [\varepsilon_2/\varepsilon_1 - (1 + \sigma)], \quad \Lambda_2 = \frac{1}{1 - \sigma} [(1 + \sigma) - (\varepsilon_2/\varepsilon_1)\sigma] \tag{8.25}$$

are both positive. We now rewrite (8.23) in the form:

$$V(s, y') = e^{-(1+\sigma)\varepsilon_1 \beta_{F0} y'} e^{-s\beta_{F0} y'} F(s)[1 + W(s, y')] \tag{8.26}$$

In (8.26), for convenience we have defined $W(s, y')$ in order to facilitate inverting the Laplace transformation:

$$W(s, y') = \left(e^{\Lambda_1 \frac{\beta_{F0}\varepsilon_1 y'}{s + \sigma}} - 1\right) + \left(e^{\Lambda_2 \frac{\beta_{F0}\varepsilon_1 y'}{s + 1}} - 1\right) + \left(e^{\Lambda_1 \frac{\beta_{F0}\varepsilon_1 y'}{s + \sigma}} - 1\right)\left(e^{\Lambda_2 \frac{\beta_{F0}\varepsilon_1 y'}{s + 1}} - 1\right)$$

$$\tag{8.27}$$

Let us use the symbol $\supset$ to denote the corresponding interpretation of the Laplace transformation of a function. From known results (Erdelyi et al. 1954, vol. 1, p. 244),

$$W(s,y') \supset w(x',y') = e^{-\sigma x'}\sqrt{\Lambda_1 \beta_{F0}\varepsilon_1 y'/x'}I_1\left(2\sqrt{\Lambda_1 \beta_{F0}\varepsilon_1 y'x'}\right)$$

$$+ e^{-x'}\sqrt{\Lambda_2 \beta_{F0}\varepsilon_1 y'/x'}I_1\left(2\sqrt{\Lambda_2 \beta_{F0}\varepsilon_1 y'x'}\right) + \int_0^{x'} e^{-\sigma\varsigma}\sqrt{\Lambda_1 \beta_{F0}\varepsilon_1 y'/\varsigma}I_1\left(2\sqrt{\Lambda_1 \beta_{F0}\varepsilon_1 y'\varsigma}\right)$$

$$e^{-(x'-\varsigma)}\sqrt{\Lambda_2 \beta_{F0}\varepsilon_1 y'/(x-\varsigma)}I_1\left(2\sqrt{\Lambda_2 \beta_{F0}\varepsilon_1 y'(x'-\varsigma)}\right)d\varsigma \qquad (8.28).$$

In (8.28), the function $I_1(\chi)$ is the Bessel function if imaginary argument χ of order one (Watson 1962, p. 77), and in the last term, use is made of the convolution theorem. Similarly, with the use of the convolution theorem, we can write for the last term in (8.26):

$$F(s)W(s,y') \supset \int_0^{x'} f(\varsigma)w(x'-\varsigma,y')d\varsigma \qquad (8.29),$$

and also the factor $e^{-s\beta_{F0}y'}$ in (8.26) lends itself to the "shift rule" (Carslaw and Jaeger 1948; Churchill 1959) so that the operational solution for $v'(x',y')$ is then:

$$V(s,y') \supset v'(x',y') = e^{-(1+\sigma)\varepsilon_1 \beta_{F0}y'}$$

$$\left[f(x'-\beta_{F0}y') + \int_0^{x'-\beta_{F0}y'} f(\varsigma)w(x'-\beta_{F0}y'-\varsigma,y')d\varsigma\right] \qquad (8.30),$$

where the function $w(x',y')$ is defined in (8.28). Along the frozen Mach wave $x' = \beta_{F0}y'$ the exponential decay of disturbance generated by the obstacle is obtained. The exponential decay factor in (8.30) is, after expanding out the definition of ε_1:

$$(1+\sigma)\left(\frac{\beta_{I0}}{\beta_{F0}} - 1\right)\beta_{F0}y' \qquad (8.31).$$

The exponential decay factor in (7.16) from the full wave equation is:

$$\frac{1+\sigma}{2}\left[\left(\frac{\beta_{I0}}{\beta_{F0}} - 1\right)\left(\frac{\beta_{I0}}{\beta_{F0}} + 1\right)\right]\beta_{F0}y' \qquad (8.32).$$

It is consistent with (8.31) of the dilute particle phase approximation $\kappa_0 \ll 1$, where for the same approximation the factor is recovered $\left(\frac{\beta_{I0}}{\beta_{F0}} + 1\right) \approx 2$.

We can interpret (8.30) as the response, in terms of the normal velocity, as the result of the superposition of a sequence of small disturbances located on the $y' = 0^+$ plane, along the x'-axis. The distribution of sources is represented by $f(x')$. The induced velocity at any point x', y' in the flow field is determined by the superposition of those disturbances situated upstream from the point $\zeta = x' - \beta_{F0}y'$ in accordance with the zone of action determined by the frozen Mach wave for supersonic Mach numbers $M_{F0} > 1$.

The behavior for the far-field wave structure is readily deduced by considering $V(s, y')$ in (8.26). The standard procedure, which will not be repeated here, is identical to that described in Section 7, except that the discussion now refers to the function $S(s)$ defined by (8.19) instead of $S_1(s)$ of (7.10). It is found here that the saddle point is again $s_0 = 0$ and that the disturbances are ultimately propagated along the lowest order, equilibrium Mach waves $x' - \beta_{E0} y' = $ constant. The function $S(s)$ is analytic at the saddle point $s_0 = 0$; its Taylor expansion about $s_0 = 0$ is:

$$-mS(s) = -\bar{a}_1 s + \bar{a}_2 s^2 - \bar{a}_3 s^3 + \ldots \tag{8.33}$$

where $m = y'/x'$ and:

$$\bar{a}_1 = \beta_{E0} y'/x'$$

$$\bar{a}_2 = \frac{1+\sigma}{\sigma} \beta_{E0} \left(1 - \frac{\beta_{I0}}{\beta_{E0}}\right) y'/x' \tag{8.34}$$

$$\bar{a}_3 = \frac{\beta_{E0}}{\sigma} \left[\left(1 - \frac{\beta_{F0}}{\beta_{E0}}\right) - \frac{(1+\sigma)^2}{\sigma} \left(1 - \frac{\beta_{I0}}{\beta_{E0}}\right) \right] y'/x'$$

Comparison with (7.25) of the corresponding coefficients in the Taylor expansion of $S_1(s)$ (Section 7) shows the consistency. For instance, when $(1 - \beta_{I0}/\beta_{E0}) \ll 1$ for $\kappa_0 \ll 1$, the coefficients $a_1 \to \bar{a}_1$. Similarly, $a_2 \to \bar{a}_2$, $a_3 \to \bar{a}_3$, ... for the dilute particle phase. Since the consistency is clear, the conclusions from Section 7 about wave structure hold for dilute-particle-phase situations as well. It will be shown that the advantage in the approximate treatment from the starting point of the simple-wave approximation to the relaxation wave equation greatly facilitates the derivation of the pressure coefficient in terms of exponential functions.

Utilizing the anticipation of the Mach wave structure from the simple-wave approximation of this section, we obtain the asymptotic behavior again from the slope of the normal velocity component as expressed by the contour integral:

$$v'_{x'}(x', y') = \frac{1}{2\pi i} \int_{L_1} sF(s) e^{x's - S(s)y'} ds \tag{8.35}$$

where $S(s)$ is now given by (8.19). The location of the main part of the disturbance and its structure is similar to that discussed in Section 7 and will thus not be repeated.

The expansion (8.33) is now used in the expression for the streamwise slope of the normal velocity (8.35). The contour of integration is chosen to pass through the origin and in this case, the path L_1 is again the imaginary axis $s = is_I$:

$$\frac{\partial v'}{\partial x'} \approx \frac{1}{2\pi} \int_{-\infty}^{+\infty} is_I F(is_I) e^{ix'[(1-\bar{a}_1)s_1+\bar{a}_3 s_I+\cdots]} e^{-x'[\bar{a}_2 s_I^2 - \cdots]} ds_I \qquad (8.36),$$

with definitions of $\bar{a}_i$ given by (8.34). For the simple wedge, $is_I F(is_I) = \alpha \beta_{F0}$, thus:

$$\frac{\partial v'}{\partial x'} \approx \frac{\alpha \beta_{F0}}{\sqrt{2\pi}} \left[\frac{1}{\sqrt{2\pi}} \int_{-\infty}^{+\infty} e^{ix'[(1-\bar{a}_1)s_1+\bar{a}_3 s_I+\cdots]} e^{-x'[\bar{a}_2 s_I^2 - \cdots]} ds_I \right]. \qquad (8.37),$$

The leading terms in (8.37) give the Fourier transform of $e^{-x'\bar{a}_2 s_I^2}$, which is a Gaussian centered about the equilibrium Mach wave as depicted in Eq. (7.27) to order $\kappa_0 \ll 1$.

$$\frac{e^{-\frac{(1-\bar{a}_1)^2 x'^2}{4\bar{a}_2 x'}}}{\sqrt{2\bar{a}_2 x'}}. \qquad (8.38).$$

9 Particle Collision with the Wall and the Normal Force

Within linearized small perturbation theory, in contrast to collisions with a blunt body at hypersonic speeds (Probstein and Fassio 1970), particle collisions with the wall have a particularly relative simple description. In the first place, all collisions are referred to the $y' = 0^+$ plane and are separable from the main problem of surface pressure. According to the linear theory, the surface pressure is proportional to the local wall deflection angle. On the other hand, the normal force per unit area due to particle-wall collisions (necessarily idealized) is a second-order effect due to its Newtonian nature and is proportional to the square of the angle between the particle streamline and the boundary at the point of impact:

$$F_N \cong \rho_{P0} u_0^2 [f(x') - \theta_P]^2, \quad \text{(Newtonian)} \qquad (9.1),$$

in which the normal momentum is given up upon collision with the wall, and:

$$F_N \cong 2\rho_{P0} u_0^2 [f(x') - \theta_P]^2, \quad \text{(elastic)} \qquad (9.2),$$

in which the particles are elastically reflected upon collision. The geometry is depicted in Figure 3. The normal collision-force coefficients are defined, respectively, as:

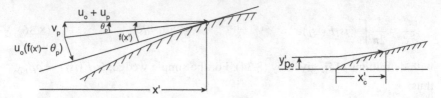

Figure 3 Particle-boundary collision in small-disturbance theory. Schematic.

$$C_{F_N, Newtonian} \equiv F_{N,Newtonian} / (\rho_0 u_0^2 / 2) \cong 2\kappa_0 \left(f(x') - \theta_P \right)^2 \qquad (9.3)$$

and

$$C_{F_N, elastic} \equiv F_{N,elastic} / (\rho_0 u_0^2 / 2) \cong 4\kappa_0 \left(f(x') - \theta_P \right)^2 \qquad (9.4).$$

The local particle streamline inclination at the wall is determined by the particle fluid interaction from the particle momentum equation (3.3), or by using (3.13) and taking the $y'-$ derivative of (3.13), which is integrated to give:

$$\theta_P \cong v_P'(x', 0) = \int_0^{x'} e^{-(x'-\xi)} f(\xi) d\xi \qquad (9.5).$$

In the foregoing, it is assumed that the obstacle has a sharp point as required by the linear theory. The post-collision particle trajectory falls within the two limiting cases from which the normal force is obtained. In the Newtonian case, the particle follows the wall after collision and forms the initial condition for post-wall-collision particle-fluid interaction or wall shape, and similarly with the opposite limiting case of elastic collision. In either case, the normal force consideration is a second-order effect. The particle trajectory up to the point of collision is:

$$y'_P = y'_{P,0} + \int_0^{x'} \theta_P(\xi) d\xi \qquad (9.6),$$

where $y'_{P,0} = y_{P,0} / \lambda_{V0}$ is the dimensionless initial normal location of the particle. The point of collision, $x'_C = x_C / \lambda_{V0}$, is then obtained implicitly from the intersection of the particle trajectory with the curve of the wall as:

$$y'_{P,0} + \int_0^{x'_C} \theta_P(\xi) d\xi = \int_0^{x'_C} f(\xi) d\xi \qquad (9.7).$$

10 The Wall Pressure Coefficient

The fluid pressure exerted on the wall is considered first from the results of the full wave equation and second from the "simple wave" approximated wave equation for dilute particle mass loading, $\kappa_0 \ll 1$. The general relations are obtained first.

We are concerned with obtaining the operational solution for the pressure coefficient:

$$C_P \equiv \frac{[p]_{y=0}}{\rho_0 u_0^2/2} \tag{10.1}.$$

The Laplace transformation of the pressure coefficient is:

$$\overline{C}_P(s) = \int_0^\infty e^{-sx'} C_P(x')dx' \tag{10.2},$$

from which the desired result is obtained from:

$$C_P(x') = \frac{1}{2\pi i} \int_{L_1} \overline{C}_P(s)e^{sx'}\,ds \tag{10.3}.$$

Applying the transformation to the momentum equations of the gas, (5.3), gives for the pressure coefficient in terms of the velocity potentials:

$$\frac{1}{2}\overline{C}_P(s) = -s\Phi(c,0) + \kappa_0[\Phi_P(s,0) - \Phi(s,0)] \tag{10.4},$$

where $\Phi(s,0)$, the transform of $\phi'(x',0)$, is:

$$\Phi(s,0) = -F(s)/S_1(s) \tag{10.5},$$

from (7.11). The momentum equation for the particle-phase (3.18), after transformation, gives:

$$\Phi_P(s,0) = \frac{1}{1+s}\Phi(s,0) \tag{10.6}.$$

Substituting (10.5) and (10.6) into (10.4), we obtain:

$$\frac{1}{2}\overline{C}_P(s) = \frac{sF(s)}{S_1(s)} + \kappa_0 \frac{sF(s)}{(1+s)S_1(s)} \tag{10.7}.$$

Recalling the definition of $S_1(s)$ given in (7.10), if we let $s_1 + s_2 = (1+\sigma)(\beta_{I0}/\beta_{F0})^2$, $s_1 s_2 = \sigma(\beta_{E0}/\beta_{F0})^2$, then we can write:

$$S_1(s) = \sqrt{\frac{(s+s_1)(s+s_2)}{(s+1)(s+\sigma)}}\beta_{F0}s \tag{10.8},$$

where

$$\frac{s_1}{s_2} = \frac{1+\sigma}{2}\left(\frac{\beta_{I0}}{\beta_{F0}}\right)^2\left[1 \pm \sqrt{1 - \frac{4\sigma}{(1+\sigma)^2}\left(\frac{\beta_{E0}}{\beta_{I0}}\right)^2\left(\frac{\beta_{F0}}{\beta_{I0}}\right)^2}\right] \quad (10.9)$$

are real quantities. It will be shown that if we obtain the pressure coefficient for a simple wedge, pressure coefficients of other general wall shapes, consistent with the linearized small perturbation theory, can be expressed in terms of the wedge through an integral representation.

10.1 The Simple Wedge of Half-Angle α

For a simple wedge of half-angle α, $F(s) = \alpha/s$, (10.7) becomes:

$$\frac{\beta_{F0}}{2\alpha}[\overline{C}_P(s)]_{wedge} = sg_1(s)g_2(s) + (1 + \kappa_0 + \sigma)g_1(s)g_2(s)$$

$$+ (1 + \kappa_0)\sigma\frac{1}{s}g_1(s)g_2(s) \quad (10.10),$$

where the shorthand notations are denoted by:

$$g_1(s) = \frac{1}{\sqrt{(s+1)(s+s_1)}} \supset e^{-\frac{s_1+1}{2}x'}I_0\left(\frac{s_1-1}{2}x'\right) \equiv G_1(x') \quad (10.11)$$

$$g_2(s) = \frac{1}{\sqrt{(s+\sigma)(s+s_2)}} \supset e^{-\frac{s_2+1}{2}x'}I_0\left(\frac{s_2-\sigma}{2}x'\right) \equiv G_2(x') \quad (10.12).$$

The Bessel function (Watson 1962, p. 77) with imaginary argument χ of order ν is denoted by $I_\nu(\chi)$. Again, the symbol $\supset$ denotes the inverse Laplace transform. The functions $G_1(x')$, $G_2(x')$ are obtained from Erdelyi et al. (1954, vol. 1, p. 235). The interpretation of the wedge pressure coefficient is then:

$$\frac{\beta_{F0}}{2\alpha}[C_P(x')]_{wedge} = \frac{dY(x')}{dx'} + (1 + \kappa_0 + \sigma)Y(x') + (1 + \kappa_0)\sigma\int_0^{x'} Y(\xi)d\xi$$

$$(10.13),$$

where the function $Y(x')$ and it's derivative are:

$$Y(x') = e^{-\frac{s_2+\sigma}{2}x'}\int_0^{x'} e^{\frac{s_2+\sigma-s_1-1}{2}\xi}I_0\left(\frac{s_1-1}{2}\xi\right)I_0\left(\frac{s_2-\sigma}{2}(x'-\xi)\right)d\xi \quad (10.14),$$

and:

$$\frac{dY(x')}{dx'} = e^{-\frac{s_2+\sigma}{2}x'} I_0\left(\frac{s_2-\sigma}{2}x'\right)$$

$$-\frac{s_1+1}{2} e^{-\frac{s_2+\sigma}{2}x'} \int_0^{x'} e^{\frac{s_2+\sigma-s_1-1}{2}\xi} I_0\left(\frac{s_1-1}{2}\xi\right) I_0\left(\frac{s_2-\sigma}{2}(x'-\xi)\right) d\xi$$

$$+\frac{s_1-1}{2} e^{-\frac{s_2+\sigma}{2}x'} \int_0^{x'} e^{\frac{s_2+\sigma-s_1-1}{2}\xi} I_1\left(\frac{s_1-1}{2}\xi\right) I_0\left(\frac{s_2-\sigma}{2}(x'-\xi)\right) d\xi \qquad (10.15).$$

At $x' = 0^+$ a "frozen jump" from the free-stream value of zero to unity is obtained. When $x' \to \infty$, the contribution comes from the integral in the last term of (10.13), which is:

$$\frac{\beta_{F0}}{2a}[C_P(\infty)]_{wedge} = (1+\kappa_0)\sigma \int_0^\infty e^{-\frac{s_2+\sigma}{2}\xi}\left[\int_0^\xi e^{\frac{s_2+\sigma-s_1-1}{2}\varsigma} I_0\left(\frac{s_1-1}{2}\varsigma\right)\right.$$

$$\left. I_0\left(\frac{s_2-\sigma}{2}(\xi-\varsigma)\right)d\varsigma\right]d\xi \qquad (10.16).$$

The first definite integral in (10.16) can be interpreted as the Laplace transform of the convolution integral in the bracket from the $\xi-$ plane to the $(s_2+\sigma)/2-$ plane, which is rewritten as:

$$\frac{\beta_{F0}}{2a}[C_P(\infty)]_{wedge} = (1+\kappa_0)\sigma \int_0^\infty e^{-\frac{s_2+\sigma}{2}x}\left[e^{\frac{s_2+\sigma-s_1-1}{2}\xi} I_0\left(\frac{s_1-1}{2}\xi\right)\right]d\xi$$

$$\bullet \int_0^\infty e^{-\frac{s_2+\sigma}{2}\xi} I_0\left(\frac{s_2-\sigma}{2}\xi\right)d\xi \qquad (10.17).$$

From Erdelyi et al. (1954) and Churchill (1959), then,

$$\frac{\beta_{F0}}{2a}[C_P(\infty)]_{wedge} = (1+\kappa_0)\sigma \frac{1}{\sqrt{s_2}\sqrt{\sigma}\sqrt{s_1}\sqrt{1}} = (1+\kappa_0)\frac{\beta_{F0}}{\beta_{E0}} \qquad (10.18),$$

after using the definitions of s_1 s_2 from (10.9). Thus the equilibrium flow pressure coefficient is recovered far downstream on the semi-infinite wedge. The factor $1+\kappa_0$ is present in (10.18) since the density ρ_0 is used in normalizing the pressure. For equilibrium flow, if we return to the normalization by using the equilibrium free-stream mixture density $(1+\kappa_0)\rho_0$, then the pressure coefficient reverts to the well-known expression in aerodynamics $[C_P]_{wedge} = 2a/\beta_{i0}$, where for the limit of frozen flow $\beta_{F0}^2 = M_{F0}^2 - 1$, and for equilibrium flow limit $\beta_{E0}^2 = M_{E0}^2 - 1$.

For numerical evaluation of the pressure coefficient (10.13), the function $Y(x')$ defined in (10.14), its derivative defined in (10.15), and its integral in terms of Bessel functions of imaginary argument must be evaluated. This is discussed in more detail in Appendix A.

10.2 Other Wall Shapes

We return to (10.7) to deduce the pressure coefficient for more general wall shapes, consistent with small perturbation theory. For the wall boundary condition with prescribed shape $f(x')$ from previous discussions, the general result can be expressed in terms of the wedge result and written in terms of Duhamel's formula (Churchill 1959):

$$\frac{\beta_{F0}}{2} C_P(x') = \frac{d}{dx'} \int_0^{x'} f(x' - \xi) \left[\frac{\beta_{F0}}{2\alpha} [C_P(\xi)]_{wedge} \right] d\xi \qquad (10.19).$$

Alternate forms of (10.19) are also available, but (10.19) is the most compact version.

10.3 The Pressure Coefficient according to the Simple-Wave Approximation

The simple-wave approximation to the full wave equation and the consequential simplified description of the Mach wave structure are discussed in Section 8. The consequent pressure coefficient is discussed in the present section. It is shown that the relations for the pressure coefficient are considerably simplified, with no sacrifice of "accuracy."

The preliminary derivation of the pressure coefficient has some common relationships with the two wave equation (full and simplified) results, except that the function $S_1(s)$ defined in (7.10) and associated with the full wave equation is replaced by $S(s)$ defined in (8.19) in the transformed pressure coefficient of (10.7):

$$\frac{1}{2}\overline{C}_P(s) = \frac{sF(s)}{S(s)} + \kappa_0 \frac{sF(s)}{(s+1)S(s)} \qquad (10.20).$$

If we let $\tilde{s}_1 + \tilde{s}_2 = (1+\sigma)\beta_{I0}/\beta_{F0}$, $\tilde{s}_1\tilde{s}_2 = \sigma\beta_{E0}/\beta_{F0}$, then the function $S(s)$ in (8.19) can be written as:

$$S(s) = \frac{(s+\tilde{s}_1)(s+\tilde{s}_2)}{(s+1)(s+\sigma)} \beta_{F0} s \qquad (10.21),$$

where

$$\frac{\widetilde{s}_1}{\widetilde{s}_2} = \frac{1+\sigma}{2}\frac{\beta_{I0}}{\beta_{F0}}\left[1 \pm \sqrt{1 - \frac{4\sigma}{(1+\sigma)^2}\frac{\beta_{E0}}{\beta_{I0}}\frac{\beta_{F0}}{\beta_{I0}}}\right] \tag{10.22}$$

are real quantities. They differ from the s_1, s_2 of (10.9) through the $\beta-$ ratios, which are squared in (10.9). Again, we discuss the simple wedge, then (10.20) can be represented by:

$$\frac{\beta_{F0}}{2\alpha}\left[\overline{C}_P(s)\right]_{wedge} = s\widetilde{g}_1(s)\widetilde{g}_2(s) + (1 + \kappa_0 + \sigma)\widetilde{g}_1(s)\widetilde{g}_2(s)$$

$$+ (1 + \kappa)\sigma\frac{1}{s}\widetilde{g}_1(s)\widetilde{g}_2(s) \tag{10.23},$$

where:

$$\widetilde{g}_1(s) = \frac{1}{s + \widetilde{s}_1} \supset e^{-\widetilde{s}_1 x'} \equiv \widetilde{G}_1(x')$$
$$\widetilde{g}_2(s) = \frac{1}{s + \widetilde{s}_2} \supset e^{-\widetilde{s}_2 x'} \equiv \widetilde{G}_2(x') \tag{10.24}$$

$$\widetilde{g}_1(s)\widetilde{g}_1(s) \supset \widetilde{Y}(x') = \int_0^{x'}\widetilde{G}_1(\xi)\widetilde{G}_1(x'-\xi)d\xi = \frac{e^{-\widetilde{s}_1 x'} - e^{-\widetilde{s}_2 x'}}{\widetilde{s}_2 - \widetilde{s}_1} \tag{10.25}$$

from known inverse transform results (Erdelyi et al. 1954). The interpretation of the pressure coefficient becomes, similarly to (10.13),

$$\frac{\beta_{F0}}{2\alpha}\left[C_P(x')\right]_{wedge} = \frac{d\widetilde{Y}(x')}{dx'} + (1 + \kappa_0 + \sigma)\widetilde{Y}(x') + (1 + \kappa_0)\sigma\int_0^{x'}\widetilde{Y}(\xi)d\xi \tag{10.26}.$$

However, in this case, the functions involved are in terms of exponentials rather than Bessel functions of imaginary argument. In terms of the function $\widetilde{Y}(x')$ of (10.25), the pressure coefficient (10.26) is now expressed in terms of the much simpler exponential functions:

$$\frac{\beta_{F0}}{2\alpha}\left[C_P(x')\right]_{wedge} = \left[e^{-\widetilde{s}_2 x'} - \widetilde{s}_1 e^{-\widetilde{s}_2 x'}\frac{e^{(\widetilde{s}_2 - \widetilde{s}_1)x'} - 1}{\widetilde{s}_2 - \widetilde{s}_1}\right]$$

$$+ (1 + \kappa_0 + \sigma)\left[e^{-\widetilde{s}_2 x'}\frac{e^{(\widetilde{s}_2 - \widetilde{s}_1)x'} - 1}{\widetilde{s}_2 - \widetilde{s}_1}\right] + (1 + \kappa_0)\sigma\left[\frac{1}{\widetilde{s}_1\widetilde{s}_2} + \frac{\widetilde{s}_1 e^{-\widetilde{s}_2 x'} - \widetilde{s}_2 e^{-\widetilde{s}_1 x'}}{\widetilde{s}_1\widetilde{s}_2(\widetilde{s}_2 - \widetilde{s}_1)}\right] \tag{10.27}.$$

For $x' \to 0^+$, the frozen jump from the free-stream zero value to unity is obtained. As $x' \to \infty$, the exponentials decay and the equilibrium pressure coefficient $(1 + \kappa_0)\sigma/\tilde{s}_1\tilde{s}_2 = (1 + \kappa_0)\beta_{F0}/\beta_{E0}$ is recovered. Equation (10.27) is then rewritten by grouping the exponentials, a form that is convenient for numerical calculations.

For general wall shape, but consistent with small perturbation theory, the pressure coefficient is expressed, as previously, in terms of the wedge pressure coefficient of (10.19).

11 Numerical Examples

The numerical examples are obtained first for a simple wedge using the full wave equation results, and subsequently from the much simpler results of the simple-wave approximation. The double-wedge solution is also shown for various extents of equilibration on the double-wedge surface.

11.1 Pressure Coefficient from Full Wave Equation

The free-stream Mach number is taken to be $M_{F0} = 1.4142$, $\beta_{F0} = 1$. Other parameter values are $\gamma = 1.40$, $\kappa_0 = 0.25$, $c_S/c_P = 1.10$, $\lambda_{V0}/\lambda_{T0} = 0.819$. The calculated parameters are then $\beta_{E0}/\beta_{F0} = 1.3098$, $\beta_{I0}/\beta_{F0} = 1.1544$; $\sigma = 1.0442$, $s_1 = 1.6152$, $s_2 = 1.1092$ (the parameters s_1, s_2 are defined in (10.6)). The function $Y(x')$ is more conveniently obtained by directly calculating the integral representations given in (A.4). The result of the numerical integration, which will be used throughout, is shown as the solid line in Figure 4. For comparison only, a three-term series approximation with Y_1, Y_2, Y_3 given by the numerical evaluation of (A.8), (A.10), and (A.11), respectively, is shown in Figure 4 as circles superimposed on the solid line. The Bessel functions used in the series are obtained from tables in Watson (1962).

The first and third terms in the pressure coefficient, (10.13), for the simple wedge involve dY/dx' and $\int_0^{x'} Y(\xi)d\xi$, respectively. It is found that they are more conveniently obtained by performing such operations numerically on $Y(x')$ directly. All three terms in the pressure coefficient (10.13) are shown individually in Figure 5 as solid lines. They are subtracted by their respective frozen values for which $\kappa_{F0} = 0$, $\sigma_F = 1$:

$$\left[\frac{dY}{dx'}\right]_F = (1 - x')e^{-x'}$$

$$[(1 + \kappa + \sigma)Y]_F = 2x'e^{-x'} \tag{11.1}$$

$$\left[(1 + \kappa)\sigma\int_0^{x'} Y(\xi)d\xi\right]_F = 1 - (1 + x')e^{-x'}$$

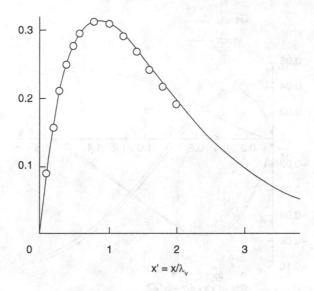

O INDICATES CALCULATED FROM SERIES (3 TERMS)

Figure 4 An integral occurring in the pressure coefficient (Appendix A) from exact treatment of small perturbation theory:
$M_0 = 1.414, \; \gamma = 1.4, \kappa_0 = 0.25, c_S/c_P = 1.1, \lambda_{V0}/\lambda_{T0} = 0.819.$ Solid line:

$$Y(x') = \int_0^{x'} e^{-\frac{s_1+1}{2}\xi} I_0\left(\frac{s_1-1}{2}\xi\right) e^{-\frac{s_2+\sigma}{2}(x'-\xi)} I_0\left(\frac{s_2-\sigma}{2}(x'-\xi)\right) d\xi$$

The sum of all three frozen values gives a value of unity. The final form of the pressure coefficient for the simple wedge is shown in Figure 6, which starts at the frozen value at the leading edge and, for sufficient streamwise distance, $x' \gg 1$, approaches the equilibrium value. The perturbation pressure coefficient $\beta_{F0} C_P/2\alpha$ at the leading edge is attributed to the abrupt turning of the gas and jumps from the zero-free-stream value to the frozen value of unity. Since the gas turns abruptly and thus much more quickly than particle velocity and thermal equilibration owing to particle phase inertia, the particles at the leading edge maintain their free-stream value. Subsequently, the relaxation processes take place and the compressed gas is accelerated and cooled by the particle phase, decreasing the pressure coefficient. Both phases then come to final equilibrium for sufficiently large streamwise distance. The pressure coefficient then corresponds to wedge flow of a single heavier gas at $[\beta_{F0} C_P/2\alpha]_E = (1 + \kappa_0)\beta_{F0}/\beta_{E0}$, since the free-stream density of the gas

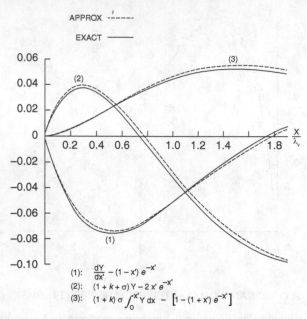

Figure 5 Integrals occurring in the pressure coefficient from exact (solid lines) and approximate (dashed lines) solution of small-disturbance theory in gas-particle flow

$$M_0 = 1.414, \ \gamma = 1.4, \kappa_0 = 0.25, c_S/c_P = 1.1, \lambda_{V0}/\lambda_{T0} = 0.819.$$

alone is used in the original normalization of the perturbation pressure, whereas in equilibrium the appropriate density is the mixture density $(1 + \kappa_0)\rho_0$.

11.2 Pressure Coefficient according to the Simple-Wave Approximation

We now turn to pressure coefficient results obtained from the relaxation wave equation to which the simple-wave approximation is made. The pressure coefficient is again obtained in terms of "Y" functions (10.25), but they are much simpler than the convolution integrals of Bessel functions, and are in terms of exponentials of the streamwise distance.

For comparison with the results from the full wave equation of Section 11.1, a similar numerical example is considered here: the numerical parameters are essentially the same, except that the new parameters are affected by the occurrence of the $\beta-$ ratios. They occur here without being squared, reflecting the

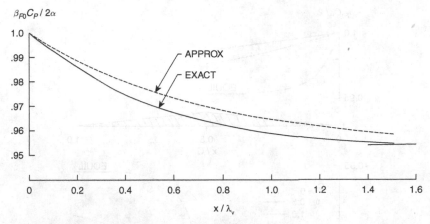

Figure 6 Pressure coefficient, $\beta_{F0}C_P/2\alpha$, as a function of distance x/λ_{V0} for a simple-wedge of angle α according to exact (solid line) and approximate (dashed line) solution of small-disturbance theory.

$$M_0 = 1.414, \quad \gamma = 1.4, \quad \kappa_0 = 0.25, \quad c_S/c_P = 1.1, \quad \lambda_{V0}/\lambda_{T0} = 0.819$$

simple-wave approximation. In this case, the new s-parameters, as defined in (10.22), are now $\tilde{s}_1 = 1.3367$, $\tilde{s}_2 = 1.0233$.

The function $\tilde{Y}(x')$ in the simple-wave approximation, given by (10.25), is shown as the dashed line in Figure 5, which compares well to the full wave equation result shown as the solid line. The remaining two functions, $d\tilde{Y}/dx'$, $\int_0^{x'} \tilde{Y}(\xi)d\xi$, are also shown as dashed lines, as indicated in Figure 5. Although the present approximate consideration requires $\kappa_0 \ll 1$, the comparison in Figure 5 for $\kappa_0 = 0.25$ does somewhat indicate that the approximation is quite "robust."

The overall pressure coefficient comparison in Figure 6 shows a difference of about 10%, or in the vicinity of an expected $\vartheta(\kappa_0^2)$ error.

11.3 The Double Wedge of Chord Length C

The results from the simple-wave approximation for the wedge of Section 11.2 are applied, where the pressure coefficient can be expressible in terms of exponential functions instead of Bessel functions. The primary aim here is to illustrate the effect of the equilibration parameter C/λ_{V0}, with $\lambda_{V0}/\lambda_{T0} = \vartheta(1)$, on the pressure coefficient (we have seen, qualitatively,

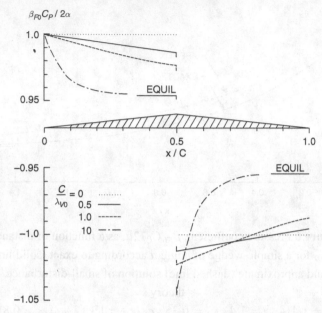

Figure 7 Pressure coefficients for double-wedge airfoil of chord length C from approximate solution of small perturbation theory for various extent of equilibration on the surface: $C/\lambda_{V0} = 0$.
dotted line, *0.5*: dark line, *1.0*: dashed line, *10*: dash-dot line
$M_0 = 1.414$, $\gamma = 1.4$, $\kappa_0 = 0.25$, $c_S/c_P = 1.1$, $\lambda_{V0}/\lambda_{T0} = 0.819$

the effect of this parameter on the far-field wave structure in Figure 2). The result for the double wedge is shown in Figure 7. For $C/\lambda_{V0} \ll 1$, the pressure coefficient very nearly corresponds to the frozen value (the doted lines) on the front part of the wedge as indicated by the solid line. For $C/\lambda_{V0} = 1$, equilibrium is not reached on the front part of the wedge. For $C/\lambda_{V0} \gg 1$, a major part of the wedge is in equilibrium flow, as indicated by the dash-dot line before the apex at the halfway point on the wedge $x/C = 0.5$ When the abrupt change in wall shape over "zero" distance is compared to λ_{V0} at mid-chord, $x/C = 0.5$, the frozen jump in pressure coefficient occurs, giving rise to gas expansion, as in ordinary gas dynamics (e.g., Liepmann and Roshko 1957). The subsequent equilibration process is similar to the front of the wedge. The extent of equilibration is always dictated by the parameter C/λ_{V0}; however, for $C/\lambda_{V0} \gg 1$, the pressure coefficient is very nearly in the equilibrium flow condition only when the surface shape is "slowly varying."

12 Relation to Other Relaxation Wave Problems

The typical relaxation wave equation, first derived by Stokes (1851) in a different context, obtained thus far indicates that Whitham's (1974) stability condition, in which the propagation speeds of the higher-order waves are greater than those of the lower-order waves, is satisfied. This is true for most relaxation wave equations.

The linearized hydrodynamic stability equation for a uniform fluidized bed (Anderson and Jackson 1969; Jackson 1971, 2000; Homsy et al. 1980), in terms of the void fraction ε, is written for normal modes in complex algebraic form. After recasting the complex frequencies into spatial and time derivatives, the hydrodynamic stability problem for a fluidized bed is interpreted as one of relaxation wave hierarchies (Liu 1981a, 1981b, 1982) for which Whitham's (1974) stability condition is violated.

The relaxation wave equation for fluidized bed instability becomes (Liu 1981a, 1981b, 1982), in the special case of one-dimensional unsteady flow,

$$\tau\left(\frac{\partial}{\partial t} + c_1\frac{\partial}{\partial x}\right)\left(\frac{\partial}{\partial t} + c_2\frac{\partial}{\partial x}\right)\varepsilon + \left(\frac{\partial}{\partial t} + a_1\frac{\partial}{\partial x}\right)\varepsilon = v_e\tau\frac{\partial}{\partial t}\frac{\partial^2}{\partial x^2}\varepsilon \qquad (12.1),$$

where τ is the relaxation time, c_1, c_2 are the higher-order wave speeds propagating in the vertical direction, a_1 is the lower-order wave speed, v_e is an effective viscosity; the wave speeds and viscosity are related to properties of the two-phase mixture in their original formulation (e.g., Jackson 1971 or Homsy, et al. 1980). Viscosity effects are stabilizing. The Whitham (1974) test for stability resides on the left side of (12.1) with respect to the relative magnitudes of the propagation speeds. The downward-propagating wave speed is given by $c_2 < 0$. There remains the comparison of the magnitude of the equivalent "frozen" higher-order wave speed c_1 and the equivalent-equilibrium wave speed a_1. In the unstable fluidized bed case, $c_1 < a_1$ and Whitham's stability condition is violated. For further discussions, see Liu (1982). This is similar to the instability of roll waves due to spillways from dams (Lighthill and Whitham 1955). Others who have contributed to the subject include (but not exhaustively) Kluwick (1983), Needham and Merkin (1983), and Ganser and Drew (1987).

The similarities between nonequilibrium reacting flow and radiating flows are discussed in Vincenti and Kruger (1965). The two sound speeds that play similar roles as the frozen and equilibrium sound speeds are the isothermal and isentropic sound speeds, which are:

$$a_T^2 = \left(\frac{\partial p}{\partial \rho}\right)_T, \quad a_S^2 = \left(\frac{\partial p}{\partial \rho}\right)_S.$$

For a perfect gas, they are $a_T^2 = RT$, $a_S^2 = \gamma RT$. The propagation of plane acoustic waves is described by (Cheng 1965; Vincenti and Kruger 1965) as:

$$\left[\left(\frac{\partial^2}{\partial x^2} - 3\alpha_0^2\right)\frac{\partial}{\partial t}\left(\frac{1}{a_{S0}^2}\frac{\partial^2}{\partial t^2} - \frac{\partial^2}{\partial x^2}\right) + \frac{16}{Bo}\alpha_0 a_{S0}\frac{\partial^2}{\partial x^2}\left(\frac{1}{a_{T0}^2}\frac{\partial^2}{\partial t^2} - \frac{\partial^2}{\partial x^2}\right)\right]\phi = 0$$

$$(12.2),$$

where α_0 is an averaged absorption coefficient having dimensions of inverse length and Bo is the Boltzman number. The respective Mach numbers are appropriately defined as:

$$M_{S0} = u_0/a_{S0}, \quad M_{T0} = u_0/a_{T0},$$

as are the Prandtl–Glauert–Ackeret parameters:

$$\beta_{T0}^2 = M_{T0}^2 - 1, \quad \beta_{S0}^2 = M_{S0}^2 - 1.$$

After the differential approximation (see Cheng 1965; Vincenti and Kruger 1965), the two-dimensional Prandtl–Glauert–Ackeret problem for a radiating gas becomes:

$$\frac{\partial}{\partial x}\left(\nabla_{xy}^2 - 3\alpha_0^2\right)\left(\beta_{S0}^2\frac{\partial^2}{\partial x^2} - \frac{\partial^2}{\partial y^2}\right)\phi + \alpha_0\frac{16}{Bo}\nabla_{xy}^2\left(\beta_{T0}^2\frac{\partial^2}{\partial x^2} - \frac{\partial^2}{\partial y^2}\right)\phi = 0$$

$$(12.3).$$

Vincenti and Kruger (1965) point out that (12.2) and (12.3) are not exactly similar in that, for the Ackeret problem, the streamwise coordinate is not a time-like variable because signals in radiative flows propagate at the speed of light and could therefore move in the negative streamwise direction opposite a supersonic stream. However, no signals can propagate in the "negative time direction."

13 Concluding Remarks

Inviscid compressible flow is studied with applications to first-order small perturbation theory. Starting from acoustic propagation in a compressible flow and particulate mixture originally at rest and in thermal equilibrium, the relaxation wave equations in airfoil coordinates are derived through a Galilean transformation, thereby connecting the acoustic and aerodynamic points of view for small perturbation theory in a gas containing small solid particles. In this case, there is no net entropy generation owing to particle-gas interactions, and velocity potentials for both phases exist in a first-order theory. The general relaxation wave equation for potential theory of supersonic flow is obtained, specialized to two-dimensional steady supersonic flow, and studied in detail, both in its Mach wave structures and in the surface pressure distribution for simple geometries. In the presence of the two relaxation processes of momentum and thermal nonequilibrium between the gas and particle phases, there are three identifiable and distinctive Mach waves: the leading and highest-order frozen Mach wave, which determines the zone of action; the lowest-order equilibrium Mach wave, which supports the bulk of the disturbance propagation into the far field with a diffusive structure; and an intermediate Mach wave, which depends on the relative importance of momentum and thermal equilibration. Signals along this wave are damped, owing to the influence of the frozen wave, and diffusive, owing to the lower-order equilibrium wave. Analytical representations are obtained for these structures through asymptotic representations. Exact representations are obtained for the pressure coefficient on the disturbance surface. For the simple wedge, for instance, the results show the transition from a frozen jump at the leading edge toward equilibrium flow far downstream. For a body of finite chord length C, the magnitude of C/λ_{V0}, C/λ_{T0} indicates to what extent equilibration takes place on the body surface.

For two-dimensional supersonic flow, or planar problems in a quasi-two dimensional supersonic flow, a simplification is possible for $\kappa_0 \ll 1$. In this situation, the Mach wave structures emanating from a single point, though distinct, are clustered closely together. The approximations due to Whitham are applied, retaining only waves inclined to the downstream direction, or a simple-wave approximation is applied directly to the full wave equation. Concurrently, approximations of a boundary-layer nature could also be applied to the full relaxation wave equation through strained coordinate transformations. These two approximations lead to the same resulting simple-wave relaxation wave equation. While the Mach wave structures deduced are consistent with the results from the full wave equation, the wall pressure coefficient is enormously simplified and is thus conveniently used to construct the pressure coefficient on a double wedge in supersonic flow.

The information about the extent of equilibration on the wall is inherently fed into the far field via the equilibrium Mach waves, and this is used to explain the transition of the shape of an oblique shock wave starting from a frozen shock angle, transitioning to an equilibrium shock angle in the far field from the present considerations. These results are deemed indispensable in explaining and understanding numerical computational results.

Finally, estimation is given in Appendix B of gas-particle parameters relatable to Martian Atmospheric Dust.

Symbols

a	speed of sound
c_P, c_V	specific heats of gas
c_S	specific heats of solid material
C	chord length
C_P	pressure coefficient
D_{12}	an effective diffusivity
e	internal energy per unit mass
$f(x')$	local surface inclination
F_P	force per unit volume exerted by particle on gas
h	static enthalpy per unit mass
I_ν	Bessel function with imaginary argument of order ν
k	thermal conductivity of gas
m_P	particle mass
M	Mach number
n_P	number density of particles
p	gas pressure
p_0	free-stream gas pressure
$p' = p - p_0$	perturbation pressure
$\mathrm{Pr} = c_P\mu/k$	gas Prandtl number
Q_P	heat transfer per unit volume from particles to gas
r_P	average particle radius
R	specific gas constant
u_0	free-stream velocity
u_j	velocity components
x_j	Cartesian coordinates

Greek Symbols

$\theta_P(x')$	local particle streamline inclination at wall
$\kappa = \rho_P/\rho$	particle mass loading
λ	particle equilibration length
μ	gas viscosity
ν	gas kinematic viscosity
$\xi, \ \eta = (x' - \beta_{F0}y')/\beta_{F0}^2, \ y'/\beta_{F0}$	
ρ	gas density

ρ_0 free-stream gas density

$\rho' = \rho - \rho_0$ perturbation gas density

$\sigma = (\lambda_{V0}/\lambda_{T0})(1 + \kappa_0 c_S/c_P)$

τ	particle equilibration time
τ_{ij}	gas viscous stress tensor
ϕ	perturbation velocity potential
$\phi' = \phi/\lambda_{V0}u_0$	dimensionless velocity potential
$\Phi'(s, y')$	Laplace transform of $\phi'(x', y')$
$\Phi_S(x, y, z)$	steady part of perturbation velocity potential
Φ	gas dissipation function
Φ_P	rate of work done, per unit volume, by particles on gas

Subscripts

0	free stream
E	equilibrium
F	frozen
P	particle
S	solid
T	temperature
V	velocity

Other less frequently used symbols are defined where they first appear.

Appendix A

Convolution Integral of Two Bessel Functions with Imaginary Argument

The integral that needs to be evaluated in the exact form of the pressure coefficient for a wedge is:

$$Y(x') = e^{-\frac{s_2+\sigma}{2}x'} \int_0^{x'} e^{\frac{s_2+\sigma-s_1-1}{2}\xi} I_0\left(\frac{s_1-1}{2}\xi\right) I_0\left(\frac{s_2-\sigma}{2}(x'-\xi)\right) d\xi \qquad (10.14).$$

In "shorthand" notation, the integral is recast into the form:

$$Y(x') = \int_0^{x'} e^{-a_1'\varsigma} I_0(b_1'\varsigma) e^{-a_2'(x'-\varsigma)} I_0(b_2'(x'-\varsigma)) d\varsigma \qquad (A.1),$$

where:

$$a_1' = (s_1+1)/2, \; a_2' = (s_2+\sigma)/2, \; b_1' = (s_1-1)/2, \; b_2' = (s_2-\sigma)/2 \qquad (A.2).$$

The integral (A.1) does not appear in readily available integrals of products of Bessel functions (e.g., Luke 1962). To proceed, we follow Erdelyi et al. (1954): the two Bessel functions with imaginary arguments are replaced by their respective Poisson's integral representation:

$$I_v(z) = \frac{(z/2)^v/2}{\Gamma(v+1/2)\Gamma(1/2)} \int_0^\pi \left(e^{z\cos\theta} - e^{-z\cos\theta}\right) \sin^{2v}\theta \, d\theta \qquad (A.3).$$

Changing the order of integration and carrying out the integration involving x' first gives:

$$Y(x') = \frac{1}{(2\pi)^2} \int_0^\pi \int_0^\pi \frac{e^{(-a_1'+a_2'+b_1'\cos\theta-b_2'\cos\bar\theta)x'} - 1}{-a_1'+a_2'+b_1'\cos\theta-b_2'\cos\bar\theta} e^{(-a_2'+b_2'\cos\bar\theta)x'} d\theta d\bar\theta$$

$$+ \frac{1}{(2\pi)^2} \int_0^\pi \int_0^\pi \frac{e^{(-a_1'+a_2'-b_1'\cos\theta-b_2'\cos\bar\theta)x'} - 1}{-a_1'+a_2'-b_1'\cos\theta-b_2'\cos\bar\theta} e^{(-a_2'+b_2'\cos\bar\theta)x'} d\theta d\bar\theta$$

$$+ \frac{1}{(2\pi)^2} \int_0^\pi \int_0^\pi \frac{e^{(-a_1'+a_2'+b_1'\cos\theta+b_2'\cos\bar\theta)x'} - 1}{-a_1'+a_2'+b_1'\cos\theta+b_2'\cos\bar\theta} e^{(-a_2'-b_2'\cos\bar\theta)x'} d\theta d\bar\theta$$

$$+ \frac{1}{(2\pi)^2} \int_0^\pi \int_0^\pi \frac{e^{(-a_1'+a_2'-b_1'\cos\theta+b_2'\cos\bar\theta)x'} - 1}{-a_1'+a_2'-b_1'\cos\theta+b_2'\cos\bar\theta} e^{(-a_2'-b_2'\cos\bar\theta)x'} d\theta d\bar\theta$$

$$(A.4).$$

The exponential function in the integrals of (A.4) can be written as:

$$\frac{1}{\eta}\left(e^{\eta x'} - 1\right) = \frac{1}{\eta}\left[\eta x' + \frac{(\eta x')^2}{2!} + \frac{(\eta x')^3}{3!} + \cdots\right] = \sum_{m=1}^{\infty} \eta^{m-1} \frac{(x')^m}{m!} \quad (A.5);$$

with radius of convergence of the series is unlimited. The integrals of (A.4) can now be written with the use of (A.5):

$$Y(x') = \frac{e^{-a_2' x'}}{(2\pi)^2} \sum_{m=1}^{\infty} \frac{(x')^m}{m!} \int_0^\pi \int_0^\pi \left\{ \left[\left(-a_1' + a_2' + b_1'\cos\theta - b_2'\cos\bar\theta\right)^{m-1}\right.\right.$$

$$+ \left(-a_1' + a_2' - b_1'\cos\theta - b_2'\cos\bar\theta\right)^{m-1}\right] e^{\left(b_2'\cos\bar\theta\right)x'}$$

$$+ \left[\left(-a_1' + a_2' + b_1'\cos\theta + b_2'\cos\bar\theta\right)^{m-1}\right.$$

$$+ \left.\left(-a_1' + a_2' - b_1'\cos\theta + b_2'\cos\bar\theta\right)^{m-1}\right] e^{-\left(b_2'\cos\bar\theta\right)x'} \right\} d\theta d\bar\theta \quad (A.6),$$

which can be written as a series:

$$Y(x') = \sum_{m=1}^{\infty} Y_m(x') \quad (A.7).$$

For $m = 1$,

$$Y_1(x') = x' I_0(b_2 x') e^{-a_2' x'} \quad (A.8),$$

through the use the expression (A.3). For higher then $m = 1$, the following representations of Bessel functions of imaginary argument of integral order are required (Watson 1962):

$$I_n(z) = \frac{(-1)^n}{\pi} \int_0^\pi e^{-z\cos\theta} \cos n\theta d\theta, \quad I_n(z) = \frac{1}{\pi} \int_0^\pi e^{z\cos\theta} \cos n\theta d\theta \quad (A.9),$$

where n is an integer. A few higher-order terms are given in the following:

$$Y_2(x') = \frac{x'^2}{2!}\left[\left(-a_1' + a_2'\right)I_0\left(b_2' x'\right) - b_2' I_1\left(b_2' x'\right)\right] e^{-a_2' x'} \quad (A.10)$$

$$Y_3(x') = \frac{x'^3}{3!}\left\{\left[\left(-a_1' + a_2'\right)^2 + b_2'^2\right]I_0\left(b_2' x'\right)\right.$$

$$\left. - \left[2b_2'\left(-a_1' + a_2'\right)\right]I_1\left(b_2' x'\right) + b_2'^2 I_2\left(b_2' x'\right)\right\} e^{-a_2' x'} \quad (A.11)$$

$$Y_4(x') = \frac{x'^4}{4!}\left\{\left[\left(-a_1' + a_2'\right)^3 + \frac{3}{2}\left(-a_1' + a_2'\right)\left(b_1'^2 - b_2'^2\right)\right]I_0\left(b_2' x'\right)\right.$$

$$- \left[3b_2'\left(-a_1' + a_2'\right)^2 + \frac{3}{2}b_1'^2 b_2' + \frac{3}{4}b_2'^3\right]I_1\left(b_2' x'\right)$$

$$+ \left[\frac{3}{2} \left(-a_1' + a_2' \right) b_2'^2 \right] I_2 \left(b_2' x' \right) - \frac{1}{4} b_3'^3 I_3 \left(b_2' x' \right) \right\} e^{-a_2' x'} \qquad \text{(A.12)}$$

Generalizations to convolution integrals when the Bessel functions are of higher order can be made easily using procedures similar to those described here.

It is sufficient to consider the function $Y(x')$ alone as it is more convenient to obtain it's derivative and integral numerically from the function itself.

Appendix B
Estimation of Gas-Particle Parameters of Martian Atmospheric Dust

The issue of Martian atmospheric dust is important in the planning and possible execution of human exploration of Mars. Recent knowledge about Martian atmosphere dust properties comes mainly from remote sensing via optical measurements. The dust materials are mainly Al_2O_3, Fe_2O_3, SiO_2, (aluminum oxide, ferric oxide, and silicon dioxide, respectively). The size distribution (Toon et al. 1977; Pollack et al. 1995; Tomasko et al. 1999) of dust particles peaks around $r_P \approx \vartheta(1\mu m)$. The dust number density near the Martian surface (Moroz et al. 1993) is $n_P \approx 1 - 2 cm^{-3}$ in constant haze. Metzger et al. (1999) found the dust mass density for standard conditions to be $\rho_P \approx 1.8 \times 10^{-7} kg\ m^{-3}$ and $\rho_P \approx 7 \times 10^{-5} kg\ m^{-3}$ during a dust storm. These correspond, respectively, to number densities of $n_P \approx 2\ cm^{-3}$ and $n_P \approx 1.5 \times 10^3\ cm^{-3}$ (Esposito et al. 2011) for particles of $r_P \approx 1.5\ \mu m$ and material density $\rho_s \approx 2.6\ g\ cm^{-3}$. From these data, the important mass-loading parameter $\kappa = \rho_P/\rho$ in gas-particle flow can be estimated from $\rho_P \approx 10^{-4}\ kg\ m^{-3}$, and from NASA's Mars atmospheric model the ambient gas $\rho \approx 10^{-2}\ kg\ m^{-3}$, giving $\kappa \approx 10^{-2}$. Thus during the aftermath of a dust storm, the gas-particle flow borders on the lower end of coupled gas-particle flow. With $n_P \approx 10^3\ cm^{-3}$, a continuum description of the dust particle phase suffices if the engineering apparatus is significantly larger than the scale of one centimeter.

A question naturally arises about the electrification of the fine dust particles. Some preliminary answers might possibly be found in laboratory wind-tunnel tests. Merrison et al. (2004) and Merrison et al. (2012) report on an electric field applied to drift dust particles out of suspension and collected on the electrodes. The mass of dust collected is then measured optically as a function of the applied electric field to determine an average electrification per grain. In the absence of large sand grains, the net electrification of micron-sized dust particles is close to zero, and is thus not expected to generate electric fields. However, if the dust particles agglomerate into effectively large particles, then electrification occurs. These issues remain to be studied further in the laboratory and also by in situ measurements.

This discussion is preliminary and must be updated as new information becomes available in connecting Martian dust parameters to gas-particle flows.

References

Ackeret, J. 1925. Luftkrafte auf Flugel, die mit grosserer als Schallgeschwindigkeit bewegt warden. *Z. Flugtech. u. Motorluftschiffahrt* **16**, 72–74 (Transl. in NACA Tech. Memo. 317).

Anderson, T. B., and Jackson, R. 1969. A fluid mechanical description of fluidized beds: Comparison of theory and experiment. *Ind. Engng. Chem. Fundam.* **8**, 137–144.

Brenan, C. E. 2005. *Fundamentals of Multiphase Flow.* Cambridge: Cambridge University Press.

Busemann, A. 1935. Aerodynamischer Auftrieb Uberschallgeschwindigkeit. In *Atti del Convegno della Fundazione Alessandro Volta.* pp. 328–360. (Transl. in British R, T.P. Translation No. 2844).

Carrier, G. F. 1953. Boundary layer problems in applied mechanics. *Advances in Applied Mechanics*, vol. **3**, 1–19 (R. von Mises and Th. von Kármán, eds.). New York: Academic Press.

Carrier, G. F. 1954. Boundary layer problems in applied mathematics. *Comm. Pure Appl. Math.* **7**, 11–17.

Carrier, G. F. 1958. Shock waves in a dusty gas. *J. Fluid Mech.* **4**, 376–382.

Carslaw, H. S., and Jaeger, J. C. 1948. *Operational Methods in Applied Mathematics*, 2nd ed. Oxford: Clarendon (New York: Dover, 1963).

Carslaw, H. S., and Jaeger, J. C. 1959. *Conduction of Heat in Solids*, 2nd ed. Oxford: Clarendon.

Cheng, P. 1965. Study of the flow of a radiating gas by a differential approximation. PhD dissertation, Stanford University.

Chu, B. T. 1957. Wave propagation and the method of characteristics in reacting gas mixtures with applications to hypersonic flow. Brown University, Division of Engineering Report WADC TN-57-213, AD 118350.

Chu, B. T., and Parlange, Y. 1962. A macroscopic theory of two-phase flow with mass, momentum, and energy exchange. Brown University, Division of Engineering Report DA–4761/4.

Churchill, R. V. 1959. *Operational Mathematics*, 2nd ed. New York: McGraw-Hill.

Culick, F. E. C., and Yang, V. 1992. Prediction of the stability of unsteady motions in solid propellant of rocket motors. Chapter 18 in *Nonsteady*

Burning and Combustion Stability Solid Propellants (L. DeLuca and M. Summerfeld, eds.), *Prog. Astron. Aeronau.* **143**, 719–779.

Dalla Valle, J. M. 1948. *Micrometrics*, 2nd ed. New York: Pitman.

Erdelyi, A., Magnus, W., Oberthettinger, F., and Tricomi, F. G., eds. 1954. *Tables of Integral Transforms, vol. I.* Bateman Manuscript Project, California Institute of Technology. New York: McGraw-Hill.

Esposito, F., Colangeli, L., Della Corte, V., Molfese, C., Palumbo, P., Ventura, S., Merrison, J., Nørnberg, P., Rodriguez-Gomez, J. F., Lopez-Moreno, J. J., del Moral, B., Jerónimo, J. M., Morales, R., Battistelli, E., Gueli, S., Paolinetti, R., and International MEDUSA Team. 2011. MEDUSA: Observation of atmospheric dust and water vapor close to the surface of Mars. *Mars* **6**, 1–12. (doi: 10.1555/mars.2011.0001).

Fan, L.-S., and Zhu, C. 1998. *Principles of Gas-Solid Flows.* Cambridge: Cambridge University Press.

Ganser, G. H., and Drew, D. A. 1987. Nonlinear periodic waves in a two-phase flow model. *SIAM J. Appl. Math.* **47**, 726–736.

Gelder, G. F., Smyers, W. H., and Glahn, U. von. 1956. Experimental droplet impingement on several two-dimensional air-foils with thickness ratios of 6 to 16 percent. *NACA TN* 3839.

Glauert, H. 1928. The effect of compressibility on the lift of an airfoil. *Proc. Roy. Soc. A* **118**, 113–119.

Glauert, M. 1940. A method of constructing the paths of rain-drops of different diameters moving in the neighborhood of (1) a circular cylinder, and (2) an airfoil, placed in a uniform stream of air; and the determination of deposit of drops on the surface and the percentage of drops caught. British Ministry of Supply, ARC RM 2025.

Guazzelli, E., and Morris, J. F. 2012. *A Physical Introduction to Suspension Dynamics.* Cambridge: Cambridge University Press.

Hermans, J. J. 1953. *Flow Properties of Disperse Systems.* Amsterdam: North Holland.

Hinze, J. O. 1962. Momentum and mechanical-energy balance equations for a flowing homogeneous suspension with slip between the two phases. *Appl. Sci. Res. A* **11**, 33–46.

Hoglund, R. F. 1962. Recent advances in gas-particle nozzle flows. *J. American Rocket Soc.* **32**, 662–671.

Homsy, G. M., El-Kaissy, M. M., and Didwania, A. 1980. Instability waves and the origin of bubbles in fluidized beds. Part II. Comparison with theory. *Int. J. Multiphase Flow* **6**, 305–318.

Ingra, O., and Ben-Dor, G. 1988. Dusty shock waves. *Appl. Mech. Rev.* **41**, 378–437.

Jackson, R. 1971. Fluid mechanical theory. In *Fluidization* (J. F. Davidson and D. Harrison, eds.). London: Academic Press.

Jackson, R. 2000. *The Dynamics of Fluidized Particles*. Cambridge: Cambridge University Press.

Kármán, Th. von 1935. The problem of resistance in compressible fluids. In *Atti del Convegno della Fundazione Alessandro Volta*. 222–277. Also in *Collected Works* **2**, 179–221. London: Butterworth (1956).

Kármán, Th. von 1941. Compressibility effects in aerodynamics. *J. Aeron. Sci.* **8**, 337–356.

Kármán, Th. von 1947a. Sand ripples in the desert. *Technicon Yearbook*, 52–54. Also in *Collected Works* **3**, 352–356. London: Butterworth (1956).

Kármán, Th. von 1947b. Supersonic aerodynamics: Principles and applications. *J. Aeron. Sci.* **14**, 373–409. Also in *Collected Works*. London: Butterworth (1956).

Kármán, Th. von 1959. Some significance developments in aerodynamics since 1946. *J. Aero/Space Sci.* **26**, 129–144. Also in *Collected Works*. London: Butterworth (1956).

Kármán, Th. Von, and Moore, N. B. 1932. Resistance of slender bodies moving with supersonic velocities. *Trans .A. S. M. E.* **54**, 303–310. Also in *Collected Works* **2**, 376–393. London: Butterworth (1956).

Kiely, D. H. 1959. The irreversible thermodynamics of particulate systems. Eng. D. thesis, Yale University.

Kluwick, A. 1983. Small-amplitude finite-rate waves in suspensions of particles in fluids. *ZAMM* **63**, 161–171. (presented at EUROMECH 1944: Mech. Sedimentation and Fluidized Beds. Tech. Univ. Vienna, 1981).

Lagerstrom, P. A. 1996. *Laminar Flow Theory*. Princeton, NJ: Princeton University Press.

Levine, J. S., and Winterhalter, D., conveners. 2017. *Dust in the Atmosphere of Mars and Its Impact on Human Exploration*. Lunar Planetary Institute, Houston. L. P. I. Contribution No. 1966, (abstracts in www.hou.usra.edu >meetings>marsdust2017).

Lewis, W., and Brun, R. J. 1956. Impingement of water droplets on a rectangular half-body in a two-dimensional incompressible flow field. NACA TN 3658.

Liepmann, H., and Roshko, A. 1957. *Elements of Gas Dynamics*. New York: Wiley.

Lighthill, M. J. 1949. A technique for rendering approximate solutions to physical problems uniformly valid. *Phil. Mag.* **40**, 1719–1201. doi: 10.1080/14786444908561410

Lighthill, M. J. 1958. *Fournier Analysis and Generalized Functions*. Cambridge: Cambridge University Press.

Lighthill, J. Sir. 1978. *Waves in Fluids*. Cambridge: Cambridge University Press.

Lighthill, M. J., and Whitham, G. B. 1955. On kinematic waves. I. Flood movement in long rivers. *Proc. Roy. Soc. A* **229**, 281–316.

Liu, J. T. C. 1964. *Problems in Particle-Fluid Mechanics*, 66–138. Ph.D.Thesis, Pasadena: California Institute of Technology.

Liu, J. T. C. 1965. On the hydrodunamic stability of a parallel dusty gas flow. *Phys. Fluids* **8**, 1939–1945.

Liu, J. T. C. 1966. Flow induced by an oscillating infinite flat plate in a dusty gas. *Phys. Fluids* **9**, 1716–1720.

Liu, J. T. C. 1967. Flow induced by the impulsive motion of an infinite flat plate in a dusty gas. *Astronautica Acta* **13**, 369–377.

Liu, J. T. C. 1981a. Finite-amplitude instabilities in fluidized beds. Abstract in *XV Biennial Symposium on Advanced Problems in Fluid Mechanics*, 80–81 (W. Ficzdon and R. Herczynski, eds.). Warsaw: IPPT-PAN.

Liu, J. T. C. 1981b. Finite-amplitude instabilities in fluidized beds. Abstract in *EUROMECH 1944: Mech. Sedimentation and Fluidized Beds*, 43–44. (R. Clift and W. Schneider, eds.). Vienna: Technical University of Vienna.

Liu, J. T. C. 1982. Note on a wave-hierarchy interpretation of fluidized bed instabilities. *Proc. R. Soc. Lond. A* **380**, 229–239.

Luke, Y. L. 1962. *Integrals of Bessel Functions*. New York: McGraw-Hill.

Marble, F. E. 1962. Dynamics of a gas containing small solid particles. In *Proc. 5th AGARD Colloquium on Combustion and Propulsion, Braunschweig*, 239–270. Oxford: Pergamon.

Marble, F. E. 1963. Nozzle contours for minimum particle-lag loss. *AIAA J.* **1**, 2793–2801.

Marble, F. E. 1969. Some gas dynamics problems in the flow of condensing vapors. *Astron. Acta* **14**, 585–613.

Marble, F. E. 1970. Dynamics of dusty gases. *Ann. Rev. Fluid Mech.* **2**, 397–446.

Merrison, J., Jensen, J., Kinch, K., Mugford, R., and Nørnberg, P. 2004. The electrical properties of Mars analogue dust. *Planet. Space Sci.* **52**, 279–290.

Merrison, J. P., Gunnlaugsson, H. P., Hogg, M. R., Jensen, M., Lykke, J. M., Bo Madsen, M., Nielsen, M. B. Nørnberg, P., Ottosen, T. A., Pedersen, R. T., Pederse, S., and Sørensen, A. V. 2012. Factors affecting the electrification of wind-driven dust studied with laboratory simulations. *Planet. Space Sci.* **60**, 328–335.

Metzger, S. M., Carr, J. R., Johnson, J. R., Parker, T. J., and Lemmon, M. T. 1999. Dust devil vortices seen by the Mars Pathfinder camera. *J. Geophys. Res. Lett.* **26**, 2781–2784.

Michael, D. H. 1964. The stability of plane Poiseuille flow of a dusty gas. *J. Fluid Mech.* **18**, 19–32.

Michael, D. H. 1965. Kelvin–Helmholtz instability of a dusty gas. *Proc. Cambridge Phil. Soc.* **61**, 569–572.

Michaelides, E. E. 2014. *Nanofluidics*. Heidelberg: Springer.

Miles, J. W. 1959. *Potential Theory of Unsteady Supersonic Flow*. Cambridge: Cambridge University Press.

Miura, H. 1974. Supersonic Flow of a Dusty Gas over a slender wedge. *J. Phys. Soc. Jpn.* **37**, 497–504.

Miura, H., and Glass, I. I. 1986 Oblique shock waves in a dusty gas flow over a wedge. *Proc. R. Soc. A* **408**, 61–78.

Miura, H., and Glass, I. I. 1988 Supersonic expansion of a dusty gas around a sharp corner. *Proc. R. Soc. A* **415**, 91–105.

Moore, F. K., and Gibson, W. E. 1960. Propagation of weak disturbances in a gas subject to relaxation effects. *J. Aero/Space Sci.* **27**, 117–127.

Moroz, V. I., Petroval, E. V., and Ksanfomality, V. 1993. Spectrophotometry of Mars in the KRFM experiment of the Phobos mission: Some properties of the particles of atmospheric aerosols and the surface. *Planetary and Space Science* **41**, 569–585.

Needham, D. J., and Merkin, J. H. 1983. The propagation of a voidage disturbance in a uniformly fluidized bed. *J. Fluid Mech.* **131**, 427–454.

Öpik, E. J. 1962. Atmosphere and surface properties of Mars and Venus. *Prog. Aeron. Scis.* **1**, 261–342. (S. F. Singer, ed.) North Holland, Amsterdam.

Othmer, D. F., ed. 1956 *Fluidization*. New York: Reinholt.

Pollack, J. B., Ockert-Ball, M. E., and Shepard, M. K. 1995. Viking lander image analysis of Martian atmosphere dust. *J. Geophys. Res.* **100** (E3), 5235–5250.

Prandtl, L. 1935 Allgemeine Uberlegungen uber die stromung zusammendruckbarer flussigkeiten. *Atti del Covegno della Foundazione Allessandro Volta*, 169–197 (Transl. in NACA Tech. Memo 805).

Prigogine, I. 1961. *Theory of Irreversible Processes*, 3rd ed. Hoboken, NJ: Wiley.

Probstein, R. F. and Fassio, F. 1970. Dusty Hypersonic Flows. *AIAA J.* **8**, 772–779. (doi: 10.2514/3.5755).

Rannie, W. D. 1962. A perturbation analysis of one-dimensional heterogeneous flow in rocket nozzles. In *Progress in Astronautics and Rocketry* **6**, 117–144 (S. S. Penner and F. A. Williams, eds.). New York: Academic Press.

Rayleigh, J. W. S., Lord. (1894) 1945. *The Theory of Sound*. New York: Dover.

Rubinow, S. I., and Keller, J. B. 1961 The transverse force on a spinning sphere moving in a viscous fluid. *J. Fluid Mech.* **11**, 447–459.

Rudinger, G. 1964. Some properties of shock relaxation in gas flows carrying small particles. *Phys. Fluids* **7**, 658–663.

Rudinger, G. 1980. *Fundamentals of Gas-Particle Flow*. New York: Elsevier.

Saffman, P. G. 1962. On the stability of laminar flow of dusty gas. *J. Fluid Mech.* **13**, 120–128.

Sears, W. R. 1954. Small perturbation theory. In *General Theory of High Speed Aerodynamics*, 61–121 (W. R. Sears, ed.). Princeton, NJ: Princeton University Press.

Serafini, J. S. 1954. Impingement of water droplets on wedges and double-wedge airfoils at supersonic speeds. NACA Report 1159.

Sneddon, I. N. 1951. *Fourier Transforms*. New York: McGraw-Hill.

Soo, S. L. 1967. *Fluid Dynamics of Multiphase Systems*. Walthham: Blaisdel.

Stokes, G. G. 1851. An examination of the possible effect of the radiation of heat on the propagation of sound. *Phil. Mag.* **1**, 305–317. Also in *Math and Phys. Papers* **3**, 142–154. Cambridge University Press (1901).

Tabakoff, W., and Hussein, M. F. 1971. Trajectories of particles suspended in fluid flow through cascades. *J. Aircraft* **8**, 60–62.

Taylor, G. I. 1940. Notes on possible equipment and technique for experiments on icing of aircraft. Britain, Ministry of Supply, ARC RM 2029.

Tomasko, M. G., Doosa, L. R., Lemmom, M., Smith, P. H., and Wegryn, E. 1999. Properties of dust in Martian atmosphere from images of Mars Pathfinder. *J. Geophys. Res.* **104**, 8987–9008.

Toon, O. B., Pollack, J. B., and Sagan, C. 1977. Mariner 9 spacecraft. Physical properties of the particles comprising the Martian dust storm of 1971–1972. *Icarus* **30**, 663–696.

Torobin, L. B., and Gauvin, W. H. 1959. Fundamental aspect of solid-gas flow. *Canadian J. Chem. Eng*: Pt. I (August 1959), 121–142; Pt. II (October 1959), 167–176; Pt. III (December 1959), 224–236; Pt. IV (October 1960), 142–153; Pt. V (December 1960) 189–200.

Van Deemter, J. J., and Van der Laan, E. T. 1961. Momentum and energy balances for dispersed two-phase flow. *Appl. Sci. Res. A* **11**, 102–108.

Vincenti, W. G. 1959. Non-equilibrium flow over a wavy wall. *J. Fluid Mech.* **6**, 481–496.

Vincenti, W. G., and Kruger, C. H., Jr. 1965. *Introduction to Physical Gas Dynamics*. New York: Wiley.

Watson, G. N. 1962. *A Treatise on the Theory of Bessel Functions*, 2nd ed. Cambridge: Cambridge University Press.

Wegener, P. P., and Cole, J. D. 1962. Experiments on propagation of weak disturbances in stationary supersonic nozzle flow of chemically reacting gas mixtures. In *Eighth Symposium (International) on Combustion*. Baltimore: Williams and Wilkins co., pp. 348–359.

Whitham, G. B. 1959. Some comments on wave propagation and shock wave structure with application of magnetohydrodynamics. *Comm. Pure Appl. Math.* **12**, 113–158.

Whitham, G. B. 1974. *Linear and Nonlinear Waves*. New York: Wiley.

Acknowledgments

I wish to thank Editors Steven Elliot, Wei Shyy, Vigor Yang, the staff at Cambridge University Press, and Katrina Avery, my lifelong editor, for their most valuable help. I am particularly grateful to Priyanka Durai (Integra Software Services), Julia Ford (Cambridge University Press) and their team for their diligent, patient and effective rendering of this Element.

Elements of Aerospace Engineering

Vigor Yang
Georgia Institute of Technology

Vigor Yang is the William R. T. Oakes Professor in the Daniel Guggenheim School of Aerospace Engineering at Georgia Tech. He is a member of the US National Academy of Engineering and a Fellow of the American Institute of Aeronautics and Astronautics (AIAA), American Society of Mechanical Engineers (ASME), Royal Aeronautical Society (RAeS), and Combustion Institute (CI). He is currently a co-editor of the Cambridge University Press Aerospace Series and co-editor of the book *Gas Turbine Emissions* (Cambridge University Press, 2013).

Wei Shyy
Hong Kong University of Science and Technology

Wei Shyy is President of Hong Kong University of Science and Technology and a Chair Professor of Mechanical and Aerospace Engineering. He is a fellow of the American Institute of Aeronautics and Astronautics (AIAA) and the American Society of Mechanical Engineers (ASME). He is currently a co-editor of the Cambridge University Press Aerospace Series, co-author of Introduction to *Flapping Wing Aerodynamics* (Cambridge University Press, 2013) and co-editor in chief of *Encyclopedia of Aerospace Engineering*, a major reference work published by Wiley-Blackwell.

About the Series

An innovative new series focusing on emerging and well-established research areas in aerospace engineering, including advanced aeromechanics, advanced structures and materials, aerospace autonomy, cyber-physical security, electric/hybrid aircraft, deep space exploration, green aerospace, hypersonics, space propulsion, and urban and regional air mobility. Elements will also cover interdisciplinary topics that will drive innovation and future product development, such as system software, and data science and artificial intelligence.

Cambridge Elements ≡

Elements of Aerospace Engineering

Elements in the Series

Distinct Aerodynamics of Insect-Scale Flight
Csaba Hefler, Chang-kwon Kang, Huihe Qiu and Wei Shyy

A Unified Computational Fluid Dynamics Framework from Rarefied to Continuum Regimes
Kun Xu

Mach Wave and Acoustical Wave Structure in Nonequilibrium Gas-Particle Flows
Joseph T. C. Liu

Printed in the United States
by Baker & Taylor Publisher Services